Understanding the Space-Time Concepts of

SPECIAL RELATIVITY

Published by
Publishers Creative Services Inc.
Distributed by
HALSTED PRESS Division of
John Wiley & Sons, Inc.
New York Toronto London Sydney

Understanding the Space-Time Concepts of SPECIAL RELATIVITY

Arthur Evett
California State University, Dominguez Hills

Understanding The Space-Time Concepts of Special Relativity

Printed in the United States of America

PUBLISHERS CREATIVE SERVICES INC.
89-31 161 Street; Suite 611
Jamaica, Queens, N.Y. 11432

Distributed by the HALSTED PRESS Division
of JOHN WILEY & SONS, INC., New York

Library of Congress Cataloging in Publication Data

Evett, Arthur.
Understanding the space-time concepts of special relativity.

Includes index.
1. Relativity (Physics) 2. Space and time.
I. Title.
QC173.65.E93 530.1'1 82-3106
ISBN 0-470-27333-X AACR2

Book Design by Sidney Solomon

With love and thanks to
Helena and Ninette,
Jeanenne and Julie

Contents

Preface

Twenty-five years of experience in teaching the concepts of relativity theory in college courses that ranged from the most elementary level for freshmen nonscience majors to the most advanced level for graduate mathematical physicists has convinced the author of the potential utility of a book such as this. It has the modest mission of presenting a *detailed*, self-contained discussion of the basic space–time concepts of special relativity, with emphasis on a thorough treatment of a variety of specific examples. Students often have difficulty in developing more than just a cursory understanding of these fundamental concepts because many of the standard texts devote very little space to their ramifications, and also because many instructors who teach courses that include some topics from relativity theory are interested more in presenting the results than in assisting the student who may desire a firmer grasp of the subtleties. It is surprising how many misconceptions are picked up by students under these circumstances.

The mathematical level has been kept as low as possible. A knowledge of elementary algebra is sufficient. Enough qualitative discussion is included so that a student whose background is weak in algebra should be able to assimilate the gist of the arguments.

Relativity theory has a natural division into two distinct stages. The first stage consists of Einstein's explanation for the strange features of the speed of light, an explanation that involves an introduction of new and novel space-time concepts. This book is concerned mainly with this crucial first stage. The last chapter outlines the approach and some of the results for the second stage, which takes up the problem of revising the laws of physics in order to make them compatible with these new space-time ideas.

SECTION I
DEVELOPMENT OF BASIC RESULTS

CHAPTER 1

The Why of Relativity

During the 20th century there have been two major upheavals in physics. And interestingly enough, peculiarities in the properties of light led to both upheavals. The strange features of the emission and absorption of light by atoms forced a drastic revision of our concepts concerning the behavior of atoms, and this revision culminated in what is known as quantum theory. The first complete quantum theory came out in the years 1925–1926.

In the late 1800s very puzzling properties about how light traveled from place to place became evident. Einstein developed his relativity theory in a successful attempt to make sense out of these properties. He published his theory in 1905.

Light was known to exhibit wavelike properties. It displayed all the usual wave phenomena, such as reflection, refraction, interference, and diffraction. Its wavelength, frequency, and wave speed could be measured. James Clerk

Maxwell, in the mid-1800s, had developed a very satisfactory mathematical theory of electricity and magnetism. Light is one form of electromagnetic radiation and Maxwell's theory was able to account for the properties of light, as well as all other electromagnetic phenomena. Maxwell's theory predicted a value for the speed of light through a vacuum that was confirmed by experiment. The symbol c is used to denote the speed of light through a vacuum, and its approximate numerical values in different systems of units are

$$\begin{aligned} c &= 186{,}000 \text{ miles per second} \\ &= 300{,}000{,}000 \text{ meters per second} \\ &= 1000 \text{ feet per microsecond} \end{aligned}$$

The speed of something ordinarily depends on who is measuring the speed. For example, suppose a train is traveling along some railroad tracks toward the right. A man in the train is walking at a speed of 3 mph (miles per hour) toward the front of the train, relative to the train. Suppose the train is moving with a speed of 60 mph relative to the tracks. If we are sitting on the ground next to the tracks, we can see the man walking toward the front of the train; see Figure 1-1.

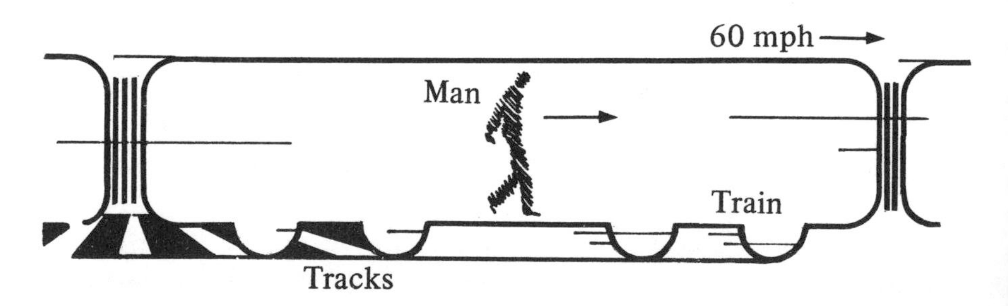

FIGURE 1-1:
Man Walking in a Moving Train

We would expect to find the speed of the man relative to us to be 63 mph, but his speed relative to observers sitting in the train would be 3 mph. In fact, we would expect the train observers to disagree with us on the speed of anything moving parallel to the tracks. This is because the train observers are moving relative to us at a speed of 60 mph. A man walking toward the *back* of the train at a speed of 3 mph relative to the train should be moving 57 mph toward the right with respect to us on the ground.

Similarly, if a beam of light were traveling toward the front of the train with a speed c relative to the train, then it should be moving (or so everyone thought) at a speed of $c + 60$ mph relative to us. It makes no sense at all, apparently, to insist that the train observers and the track observers should get the *same* value for the speed of the same light beam.

If everybody didn't get the same value for the speed of light through a vacuum, then who should get the value of c predicted by Maxwell? Many physicists gave this question much thought in the late 1800s. The nature of their deliberations is summarized in the next two paragraphs.

The "Ether"

All known waves require something material in which to propagate. In fact, each wave is a disturbance propagated through the medium that carries it. For example, a disturbance (wave) can be sent down a stretched rope by wiggling one end. Sound waves are pressure pulses (disturbances) traveling through the air. Without the air (i.e., in a vacuum) there could be no sound waves because there is nothing to disturb and hence no disturbance to propagate. So if light is a propagating disturbance (a wave), then even in a vacuum there must be something to carry the light waves. The name given to this hypothetical something was the "ether." It served the same purpose for light that air does for sound waves. When we say the speed of sound in air is 1100 ft/s (feet per second), we mean 1100 ft/s is the speed measured by an individual at rest with respect to the air that carries the sound wave. Someone moving at a speed of 100 ft/s through the air toward the sound would expect the speed of the sound to be 1200 ft/s relative to him. If he were traveling through the air at a speed of 100 ft/s away from the sound source, he would expect the speed of the sound to be 1000 ft/s relative to him. If a person were traveling through the air, he could determine how fast he was moving through the air by measuring the speed of sound relative to him coming from various directions. If he finds the speed to be the same from all directions, he must be at rest relative to the air.

By analogy, the speed c predicted by Maxwell should be the speed of light measured by an observer who happened to be at rest relative to the ether. If the observer is moving through the ether, he should find the speed of light relative to him to be different in different directions.

Difficulties with the Ether

These arguments seemed very plausible and convincing. Much effort was expended in establishing the detailed properties of the ether. It had to have the correct mechanical properties to propagate a light wave at a speed of c, yet its density had to be fantastically small because it would be present even in the best vacuum known (light was observed to travel through the nearly empty space between the stars and the earth). No one was able to construct a satisfactory theoretical model for the ether. This in itself was most frustrating.

But even worse, no one was able to detect the presence of the ether, either directly or indirectly. Its only basis for existence was that it was needed for light waves to propagate in. If an observer were moving through the ether, he should find the speed of light relative to him to be different in different directions. Experiments were designed to check this (the Michelson–Morley experiments), but the speed of light *always* seemed to be the same in all directions. The conclusion was that the earth was always at rest with respect to the ether. This was strange, because the earth travels in orbit around the sun. The earth might be at rest with respect to the ether at one time of the year, but 6 months later the earth is traveling in the opposite direction as it goes around the sun and consequently should be moving through the ether. Some scientists suggested that perhaps the ether near the earth's surface was dragged along with the earth, and therefore measurements done on the earth would always take place in this dragged-along ether. This would account for the speed of light near the earth being the same in all directions at all times.

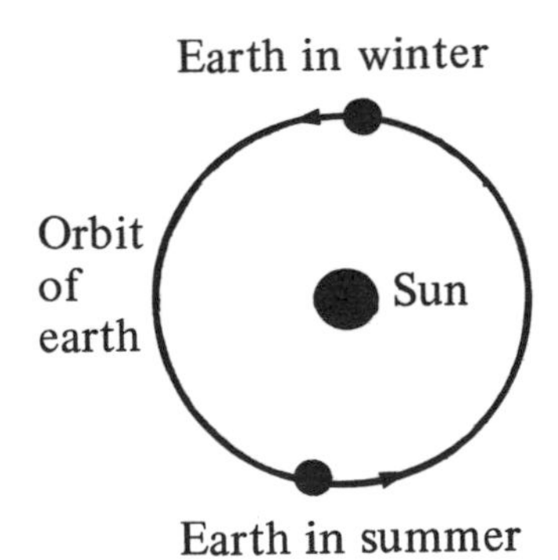

Astronomers' Objections to the Ether-Drag Idea

However, the astronomers objected strenuously to the foregoing explanation. If the ether were being dragged along in the neighborhood of the earth, the paths of light beams coming to us from the other planets and the stars would be distorted. The astronomers had allowed for no such distortion when they analyzed their observations over the centuries; yet they had been using their data to predict with great accuracy the future positions of the planets and stars and to predict eclipses. If the ether-drag idea was correct, then they would have to adjust their observations and revise all their predictions. There was already evidence for the correctness of their old predictions and these didn't allow for any ether-drag correction. Such a correction would make the revised predictions wrong. The ether-drag idea had to be discarded because it was incompatible with previous astronomical data.

This was the dilemma that faced Einstein around 1900 when he began to ponder the problem. He decided that a new fundamental approach would be necessary to resolve the difficulties, because the old ideas were leading nowhere. All the evidence seemed to indicate that the speed of light through a vacuum was the same value c in *all* directions to *all* observers, even though this didn't seem to make sense. Nevertheless, he decided to accept this property of the speed of light and then revise whatever concepts needed revising in order for it to make sense.

Einstein's Principle of Relativity

He based his approach upon what is called the "principle of relativity." One way of stating it is as follows:

> *No criterion exists for giving preference to one inertial reference frame over any other inertial reference frame.*

Let us explore the meaning of this principle in some detail. Newton introduced the idea of an inertial reference frame along with his laws of motion. An inertial reference frame is a laboratory with respect to which his laws of motion are valid. In practice, any laboratory not rotating with respect to the so-called fixed stars and moving so that an object in the laboratory with no forces on it has no acceleration relative to the laboratory is found to be an inertial frame. It can be shown that if one laboratory constitutes an inertial frame, then any other nonrotating laboratory moving with constant velocity (i.e., with constant speed in a straight line) with respect to the first laboratory is also an inertial frame.

An Old Form of the Principle of Relativity

Even in Newton's time, a *restricted* principle of relativity was thought to be valid. It could be stated as follows: No experiment based solely on Newton's laws of motion (i.e., the branch of physics known as mechanics) would yield a criterion for giving preference to one inertial reference frame over any other inertial reference frame. This merely meant that one inertial frame was as good as another insofar as Newton's laws of motion were concerned.

But suppose there is an ether. Then experiments on light performed in a room at rest with respect to the ether would give simpler results than experiments done in a room moving through the ether, for exactly the same reason that experiments on sound in air yield simpler results to an observer at rest in the air than to an observer moving through the air. For such experiments on light, one reference frame (the one at rest in the ether) would be a preferred frame.

Einstein argued that our universe is governed by a principle of relativity that claims no criterion exists for a *preferred* inertial reference frame, no matter what kind of experiment was done in the laboratory – whether the experiment involved mechanics, light, electricity and magnetism, or whatever. Thus his principle generalized the restricted form that was accepted previously.

Consequences of the Principle of Relativity

Consequences can be drawn immediately from Einstein's principle of relativity. First, all inertial observers should get exactly the same value for the speed of light through a vacuum. For if one inertial frame got a lower value than another, then they could use this as a criterion for giving preference to one frame over another. One frame could say they prefer a frame with a lower value for the speed of light, and thus establish a criterion. Remember, the relativity principle states that "no criterion exists..."; it doesn't require the criterion to be a "good" one in some absolute sense. No phenomenon should even allow a basis for an argument to get started.

Also, the speed of light through a vacuum must be the same in all directions to all inertial observers, because if it were so in one frame but not in another then it would serve as a criterion for giving preference to one frame over another. We are assuming that space is isotropic in at least one inertial frame of reference; that is, one direction in space is fundamentally the same as any other direction in space.

Reciprocity Requirement

If the observers in one inertial reference frame measure what seem to be peculiar properties about objects at rest in a second inertial reference frame, then the observers in the second inertial frame must find exactly the same peculiarities for corresponding objects at rest in the first frame. If this reciprocity feature were not satisfied, then the lack of reciprocity would be a criterion for giving preference to one inertial frame over another. We refer to this reciprocity requirement several times in the later chapters.

According to Einstein's view, the ether cannot be detected because it doesn't exist. If it doesn't exist, it cannot be used as the basis for choosing a preferred inertial reference frame. He accepted the obvious experimental fact that light could travel through a total vacuum and didn't need the troublesome ether. After Einstein successfully carried out his program of revising concepts to explain the properties of light without the use of the ether, all interest in models of the ether disappeared.

In the following chapters we explore the implications of Einstein's ideas. Many apparently well-established concepts and prejudices will be found to require modification.

CHAPTER 2

Which Event Happens First? A Disagreement.

It is most remarkable that a light beam passing through a vacuum travels at exactly the same speed from the viewpoints of *all* inertial reference frames. As mentioned in Chapter 1, such a possibility doesn't seem to be in harmony with common sense – but many phenomena in nature at one time or another have appeared to be contrary to the established "common sense." Common sense is largely a distillation of useful prejudices, and has to be adjusted if it is shown to be incompatible with observations and experiments. One must expect some of one's prejudices to be overthrown in the face of better information.

The peculiar property of the speed of light being the same to all observers forces each of us to examine our own prejudices regarding the nature of space and time. As we discover later, moving clocks will run slow, moving meter sticks will be less than a meter in length, and observers on different inertial reference frames will disagree on whether one event happens before or after a second event.

Before embarking on a quantitative discussion of these effects, it is useful to examine qualitatively the concept of simultaneity—that is, how do we check whether two events taking place at different locations occur at the same instant in time? There is no problem in doing this if the two events happen at the same place, but there is a subtle difficulty if they are separated in space.

An obvious solution would be to have observers stationed at the locations where the events will occur. If these observers had synchronized clocks, we would ask each one to note the time of any events that occur at his location. We could then compare the time readings registered by the observers in order to determine the time sequence of the spatially separated events. And, in fact, this would be a perfectly valid procedure. But how do we synchronize all these clocks at the various locations? Synchronizing two spatially separated clocks is a special case of establishing whether two events are simultaneous. If we assume we already know how to synchronize two clocks and know what properties these synchronized clocks will exhibit, then we are assuming we *already* know the solution to the simultaneity problem.

However, we had better be careful not inadvertently to incorporate one of our prejudices into the notion of simultaneity or of synchronizing two clocks. In particular, we must be sure we do not make an assumption incompatible with the peculiarities of the speed of light. The simplest way to avoid this pitfall would be to use a scheme based on these properties of light as the method for determining whether two events are simultaneous. The answers yielded by such a scheme about the nature of simultaneity would then certainly be in harmony with the puzzling property of the speed of light. And presumably any other valid scheme for determining simultaneity would agree with our special scheme.

Scheme for Determining Simultaneity

Suppose we know ahead of time that some event is going to happen at point A and another event at point B. We wish to set up an experimental arrangement to determine which event happens first. Let us put a gadget at point A that triggers a light beam that starts from A traveling toward B at the instant the event at A occurs. A similar gadget is placed at point B that triggers a light beam traveling toward A when the event at B occurs. Now our problem is to determine which light beam was triggered first. Let us locate the position exactly halfway between points A and B. Call this central location point H.

At point H we put a gadget that detects the two light beams and determines which one arrives at H first. Since each beam has exactly the same distance to travel to get to H and the speeds of the two beams are exactly the same, if the beam from A gets to H first it must have started first. This procedure converts the problem of determining whether or not two spacially separate events are simultaneous to one of determining which of two events (the arrivals of the light beams) happens first at a single location (point H). If the two beams strike H simultaneously then the events at A and B must have been simultaneous. This method does not require that the speed of light be the same in *different* inertial frames – only that the speed is independent of the direction of the light beam in all reference frames. Observers in *any* inertial frame could set up this scheme in their own reference frame, based on the fact that the speed of light is the same in all directions in their frame.

Suppose there exists a long set of straight railroad tracks at rest in our inertial reference frame. A train is traveling toward the right along these tracks at a constant speed relative to us. The train is also an inertial reference frame. Suppose two marks are made on the side of the tracks: one at point A and one at point B. Suppose the distance between these two marks is exactly the same as the length of the moving train (at least this is true according to *our* measurements). Then the front of the train will be at point B at exactly the same instant the back is passing point A, according to us. The diagram in Figure 2-1 shows the situation from our viewpoint at the instant under consideration.

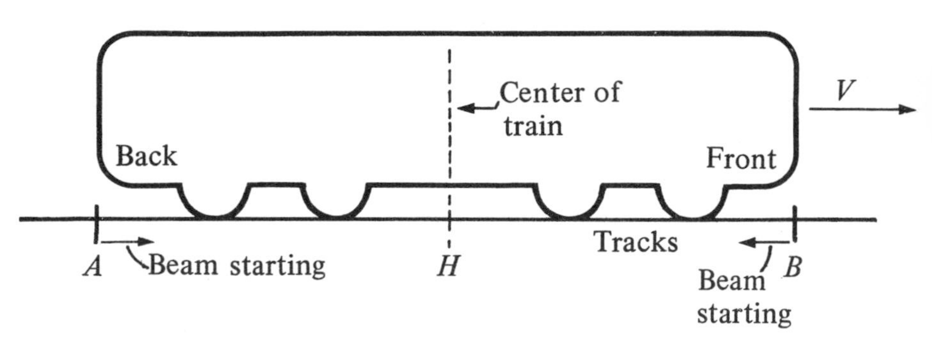

FIGURE 2-1:
Beams Leaving A and B (Our Viewpoint)

The back of the train at point A and the front of the train at point B are the two events we are comparing. These events are simultaneous to us. Consequently the light beams triggered by the events at A and B will arrive at H simultaneously. Suppose we draw the situation (from our view) at the instant the two beams get to H. By this time the train will have moved a short distance to the right. Our diagram is shown in Figure 2-2.

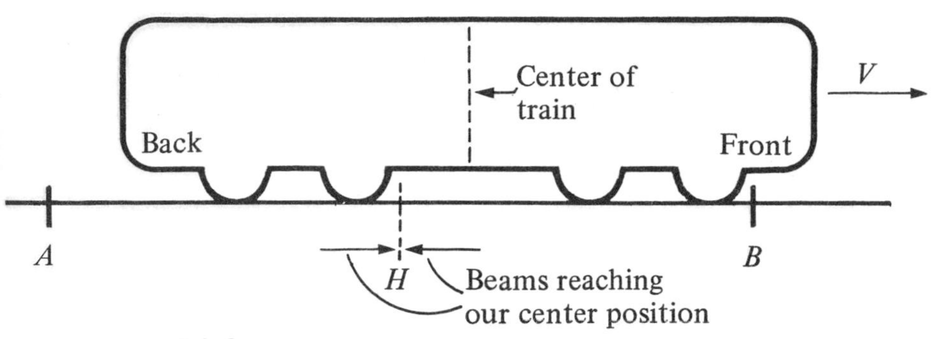

FIGURE 2-2:
Beams Arriving at H (Our Viewpoint)

Now we ask the question: Which of the two beams passed the center of the train first? It is quite clear that the one from *B* did. It has already passed the center of the train, according to Figure 2-2, and the one from *A* has not caught up to it yet. This is easy for us to explain. Because the center of the train was moving toward *B* and away from *A*, the beam from *B* had a shorter distance to travel to hit the center of the train than did the beam from *A*.

But how do the train observers account for the fact that the beam from the front of the train strikes the center of the train before the beam from the back of the train in this situation? The *only* explanation they can give is that the beam from the front must have started before the beam from the back. Remember, the center train observer is exactly halfway between the front and the back, and the speed of light in both directions is the same to the train observers. So the two events that we claim are simultaneous are *not* simultaneous from the viewpoint of the train's inertial frame. The answer to the question of whether or not two events are simultaneous is a relative one; it depends on which reference frame is asked the question. The idea that there should be only one correct answer to such a question is a prejudice nature forces us to relinquish.

Let us ask the train observers to explain the sequence of events we have described in the foregoing. We ask them to diagram the situations.

The Train's Explanation

From their view they are at rest and we are moving to the left. Also, we know from our previous discussion that they believe the front of the train is at B before the back is at A. Their diagram representing the instant the front is at B must be as shown in Figure 2-3.

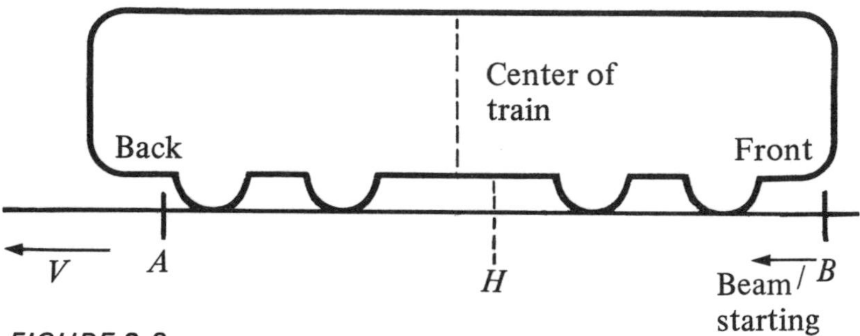

FIGURE 2-3:
Front of Train at B (Train's Viewpoint)

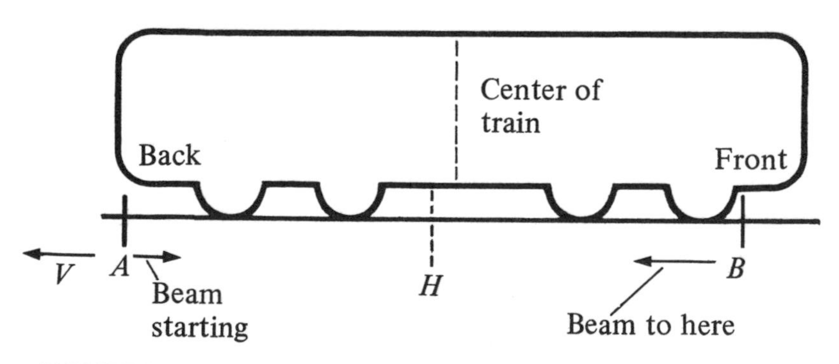

FIGURE 2-4:
Back of Train at A (Train's Viewpoint)

The train observers claim that *A* gets opposite the back of the train at a later instant, as shown in Figure 2-4.

We ask them to explain the fact that the two beams reach our *H* position at the same time. This they can easily do. According to them, our *H* position is moving toward the left, so it is traveling away from the beam from *B* and toward the beam from *A*. If the beam from *B* is to catch *H* at the same instant the one from *A* gets to *H*, then it is clear (to them) that the one from *B* has to get started first.

The train observers have a perfectly self-consistent explanation for all the phenomena involved in the process. There is no dispute over any objective piece of data. The only disagreement is about the correct explanation for the data. It is not a question of which is *the* correct explanation; the explanation given by each frame is absolutely correct from its own viewpoint. The idea that there is only one correct explanation for the time sequence of a series of events is another prejudice nature forces us to relinquish.

Inclusion of Synchronized Clocks

Let us return to the question of synchronizing a set of clocks in our reference frame. We could do this by measuring the distances between the clocks and sending light signals back and forth under prearranged conditions. For example, suppose the distance between *A* and *B* in Figure 2-1 is 1000 feet. Light travels approximately 1000 feet in 1 μs (microsecond). (For this illustration, let us suppose the speed of light is *exactly* 1000 ft/μs). It could be arranged for a light beam to be sent from *A* when the clock there reads exactly 12:00 o'clock. Then it should reach *B* when the clock at *B* reads 1 μs past 12:00 o'clock, if *B*'s clock is synchronized with *A*'s clock. If *B*'s clock didn't read the correct value, it could be reset to correct the discrepancy. By trial and error using this method, the two clocks could be synchronized. Many other schemes could be devised (using sound signals, for example), which would be equally valid.

Suppose we have synchronized clocks at A and B, and suppose the train observers have synchronized a clock at the back with another clock at the front of the train. These latter clocks are at rest relative to the train, and therefore are moving with respect to us and our clocks. Presumably we could use these pairs of clocks to determine simultaneity instead of employing the central observer described earlier in this chapter. And all the results, explanations, and conclusions should be the same as were obtained by our previously described scheme, which didn't use clocks – if a self-consistent explanation for the observations is to exist for each reference frame.

Let us include clocks in the diagram shown in Figure 2-1. For simplicity, for our clocks we will use stopwatches with only a microsecond hand. Suppose the clock at A and the clock on the back of the train happen to read straight up (zero) when they are opposite each other. Then B's clock should also read zero when the front of the train is opposite B, because the two events are simultaneous from our view. But the front of the train opposite B happened earlier than the back of the train opposite A, according to the train observers. So the clock on the front of the train would read earlier than zero (i.e., would not yet be straight up) when it is opposite B. All this is shown in Figure 2-5, which repeats Figure 2-1 with the addition of the four clocks as shown.

Let us put clocks in Figure 2-3, which is one of the train observers' diagrams. The front of the train is opposite B, so the two clocks there read the same as in Figure 2-5. The clock at the back of the train must read the same as the clock at the front of the train, because the diagram shows the situation at a given instant from the train's viewpoint. The clock at A must not yet be straight up, because it will read straight up later when it gets to the back of the train. Figure 2-6 shows the clock readings for the situation in Figure 2-3 (remember, this is drawn from the train's viewpoint).

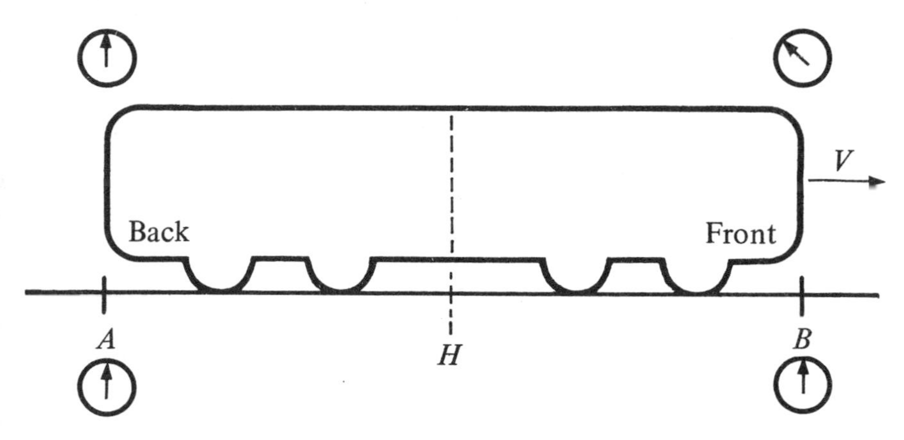

FIGURE 2-5:
Comparisons of Clock Readings (Our Viewpoint)

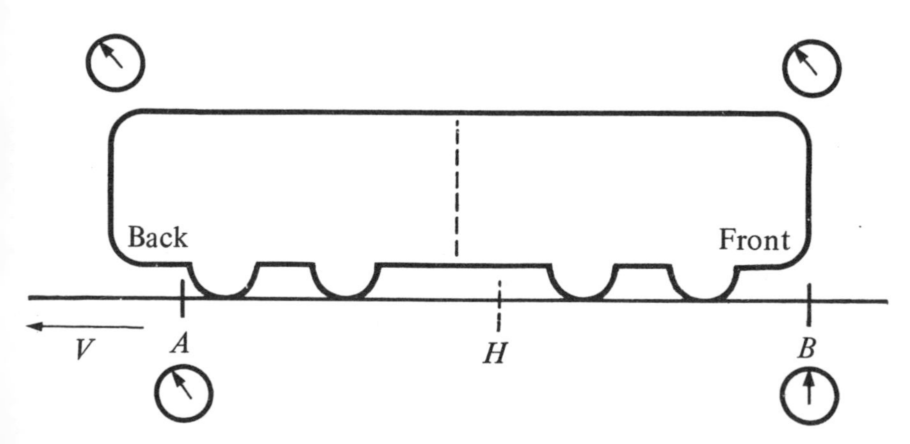

FIGURE 2-6:
Front of Train at B (Train's Viewpoint)

As shown in Figure 2-6, the train observers believe our clocks are out of synchronization, which is also our complaint about the train's clocks. In fact, each of us says the other frame's clocks are out of synchronization. The train is moving to the right with respect to us, and the clocks on the train farther right read earlier, according to us (see Figure 2-5). Similarly, we are moving toward the left with respect to the train and our clocks farther left read earlier, according to the train observers (see Figure 2-6).

Finally, let us show the clocks in the situation represented in Figure 2-4. Remember, this also is drawn from the train's viewpoint. Clock A now reads zero and clock B reads later than zero, as indicated in Figure 2-7.

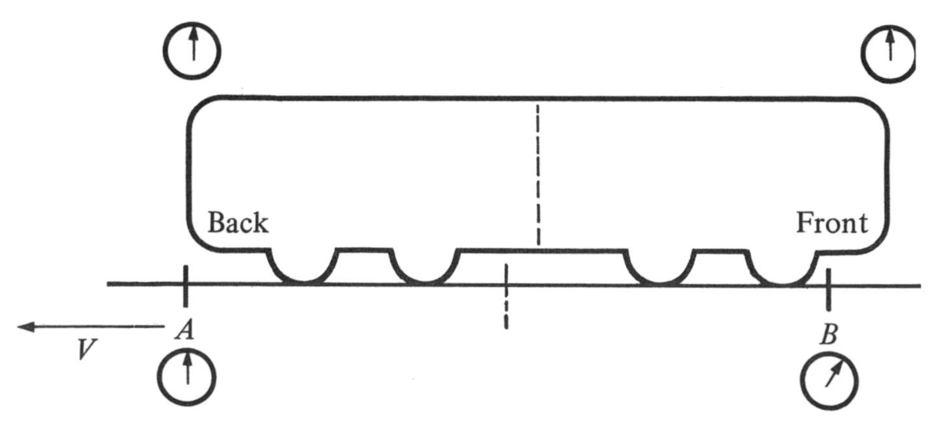

FIGURE 2-7:
Back of Train at A (Train's Viewpoint)

Disagreement on Relative Lengths

Let us consider the question: How does the distance between points A and B compare with the length of the train?

From Figure 2-1 or Figure 2-5, it is obvious that our answer is that the moving train is exactly the same length as the distance between A and B (this was part of the information used to construct Figure 2-1). Now refer to Figure 2-3, which is the train observers' diagram. Quite clearly they would answer that the train length is greater than the distance between A and B.

Thus, in addition to disagreeing on simultaneity, the two reference frames will disagree on the relative lengths of objects in certain circumstances. Another prejudice must bite the dust!

Study the preceding material carefully. Digesting the arguments that led to all the qualitative results shown in this chapter is the crucial step in understanding relativity. In Chapter 4 we analyze quantitatively the disagreement on relative lengths, and in Chapter 5 we return to a more detailed examination of simultaneity.

Suggested practice problems to ponder in Appendix A: P-1; P-2(a).

CHAPTER 3

Moving Clocks Run Slow

In Chapter 2 we used the properties of light to set up a procedure for determining simultaneity. In this way we would know that our results on the nature of simultaneity would be compatible with the properties of light. Now we wish to explore some properties of moving clocks versus stationary clocks in a manner compatible with the properties of light. To ensure this compatibility, we design our clocks using a light beam. Later we assume for consistency that *all* valid clocks have similar properties.

Our clock consists of two stationary mirrors that reflect a beam of light back and forth. An indicator hand on a meter at the bottom mirror is set to register an amount $2L/c$ every time the bouncing beam hits the bottom mirror. If L is the separation of the mirrors, then $2L/c$ is the round-trip time of the beam. Thus the light clock will correctly indicate time intervals, if constructed this way. Any other type of clock could be calibrated against this light clock.

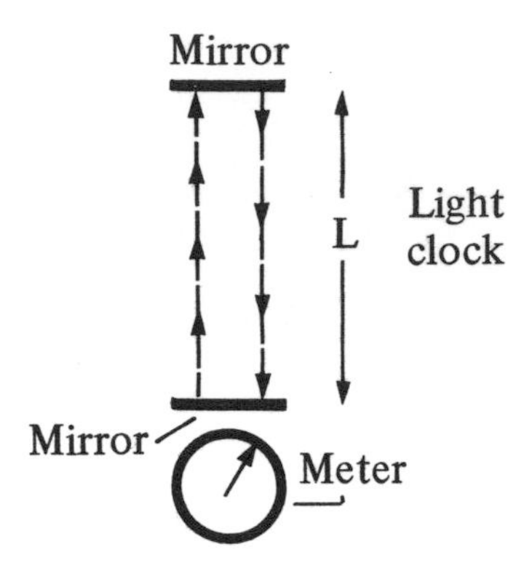

Suppose our reference frame has a set of these clocks at various locations alongside our railroad tracks. We synchronize them. The train comes by at speed V, and suppose we are interested in comparing the rate at which the train's clock runs with our own. The train observers have constructed a clock in exactly the same way we have. It registers an amount $2L/c$ after each round trip. Suppose the train clock reads zero at the instant it passes one of my clocks, which also reads zero. This initial situation is shown in Figure 3-1, from our viewpoint. Note that we claim the spacing of the mirrors on their clock is also L. This needs some justification, since we found in Chapter 2 that for some situations the two reference frames will disagree on separations.

Agreement on Separations Measured along a Line Perpendicular to the Direction of Relative Motion

The separation involved here is a rest separation from the train's viewpoint but a moving separation from our viewpoint. Nevertheless, for a separation measured along a line perpendicular to the direction of relative motion of the two frames, there must be no disagreement on the sepa-

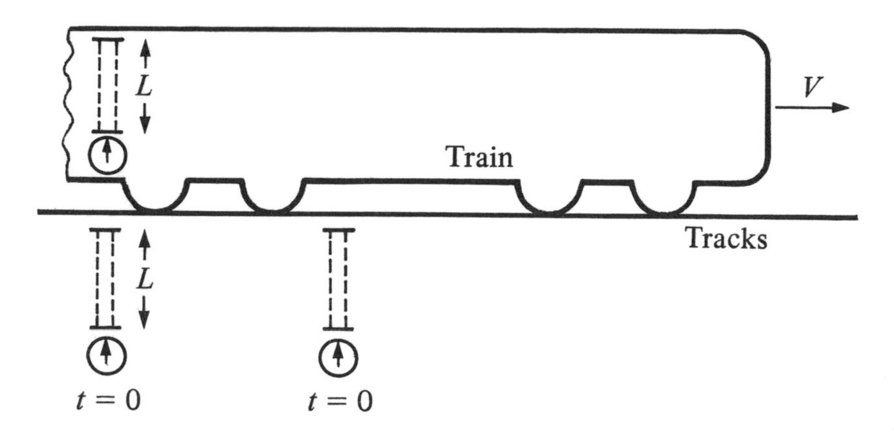

FIGURE 3-1:
Initial Situation (Our Viewpoint)

ration. This can be seen in the following way. Suppose two objects in the train are placed a distance L apart along a direction perpendicular to the relative velocity vector, according to the train's frame. But suppose this separation is less than L to us. We then could position two objects at rest a distance L apart, according to us, such that when the two moving objects in the train came by, they would both pass between our two objects. This experiment would justify our claim that separations the train observers think are L are less than L from our viewpoint. But if they interpreted the same experiment, they would have to conclude our L lengths are *greater* than theirs. This would destroy the reciprocity between the two frames, and therefore would be in violation of the principle of relativity. The only way such a difficulty can be avoided is for us to agree on separations measured along a line perpendicular to the direction of relative motion.

Let us return to Figure 3-1. Because the train's clock is moving to the right with speed V, its beam of light takes a zigzag path according to us. Let us draw the situation at

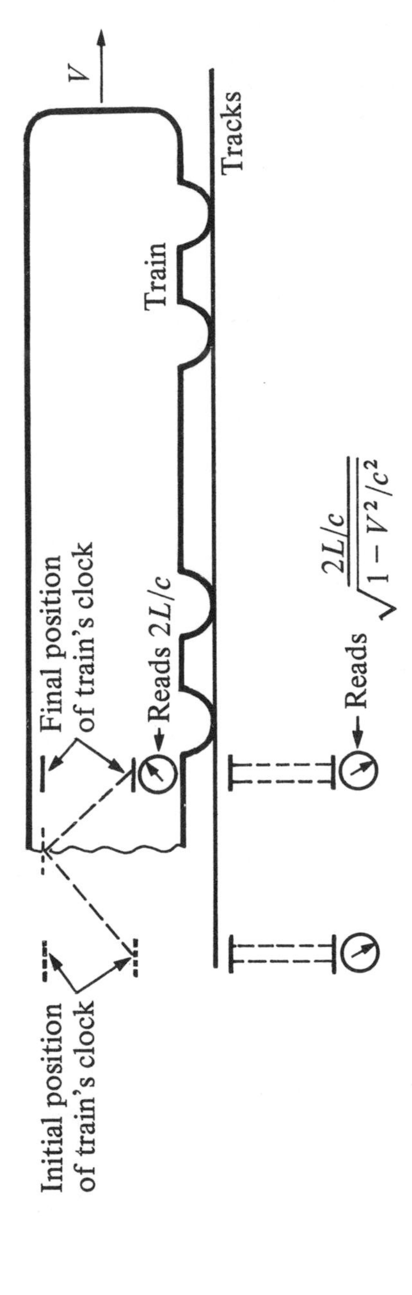

FIGURE 3–2:
Final Situation (Our Viewpoint)

the end of one round trip of the beam of light in the train's clock. The train's clock will read $2L/c$ since the round trip has been completed. But the path of the beam is longer than $2L$ according to us. Suppose the time that elapsed during the round trip, according to us, is Δt. The distance moved down the tracks by the train's clock is $V\Delta t$. So the path of the light beam leads to the triangle shown in the following diagram.

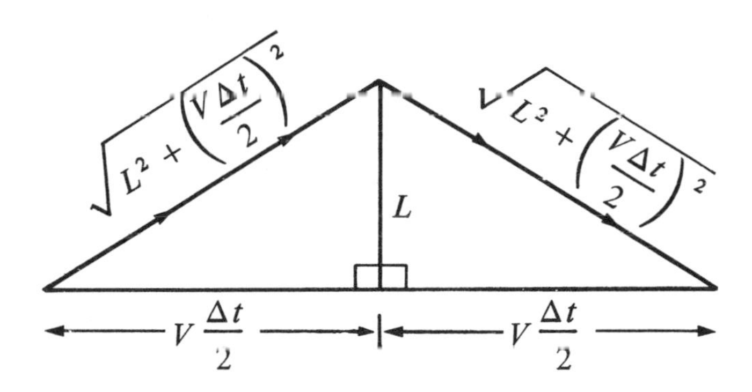

From the Pythagorean theorem we get the length of the path for the round trip to be

$$2\sqrt{L^2 + \frac{V^2\Delta t^2}{4}}\,.$$

Thus the time interval for the round trip, from our viewpoint, is

$$\Delta t = \frac{\text{Distance}}{c}\,,$$

or

$$\Delta t = \frac{2\sqrt{L^2 + V^2\Delta t^2/4}}{c}\,.$$

Let us solve for Δt. Square both sides:

$$(\Delta t)^2 = \frac{4(L^2 + V^2/4\,[\Delta t]^2)}{c^2},$$

or

$$\Delta t^2 = \frac{4L^2}{c^2} + \frac{V^2(\Delta t)^2}{c^2}.$$

Juggling this last expression gives

$$(\Delta t)^2(1 - V^2/c^2) = 4L^2/c^2,$$

or

$$(\Delta t)^2 = \frac{4L^2/c^2}{1 - V^2/c^2}.$$

Now take the square root of both sides. Finally,

$$\Delta t = \frac{2L/c}{\sqrt{1 - V^2/c^2}}$$

But $2L/c$ is the amount of time that would be registered on the moving clock. Δt is the time we say it really took for the round trip. We have derived the following relation:

$$\left(\begin{array}{c}\text{Time interval registered}\\ \text{on our clocks}\end{array}\right) = \frac{\left(\begin{array}{c}\text{Time interval registered on}\\ \text{the moving clock}\end{array}\right)}{\sqrt{1 - V^2/c^2}}$$

Multiply both sides by $\sqrt{1 - V^2/c^2}$. We get

$$\left(\begin{array}{c}\text{Time interval registered}\\ \text{on the moving clock}\end{array}\right) = \left(\begin{array}{c}\text{Time interval registered}\\ \text{on our clocks}\end{array}\right)\sqrt{1 - V^2/c^2}$$

Moving Clocks Run Slow

If V isn't zero, then $\sqrt{1 - V^2/c^2}$ is less than one, and becomes zero when $V = c$. Thus the time interval on the moving clock is less than the time interval measured on our clocks. We say "moving clocks run slow." This phenomenon is known as "time dilation." Note that in our experiment we used only one of the train's clocks but at least two of our synchronized clocks. Suppose the train observers decide to check the rate of one of our clocks. They will use two of their synchronized (to them) clocks and watch one of our clocks. They will be able to use the same argument we did to conclude that *our* clock is running slow compared with theirs. So each says the other frame's clocks are running slow by the $\sqrt{1 - V^2/c^2}$ factor. This had better be so if our results are to be in agreement with the principle of relativity. If we said their clocks run slow and they said ours ran fast, this would be a criterion for us to give preference to one inertial reference frame over another. We could say ours is best because we prefer a frame with the faster clocks. But if *both* frames find the other frame's clocks are slow, neither has any claim to being a preferred frame. This is another example of the reciprocity that must exist between any two inertial reference frames if the principle of relativity is to be satisfied.

If the train observers were watching *our* experiment, they would expect us to come to some strange conclusions about the relative rates of the two frame's clocks because, from their view, we aren't using truly synchronized clocks in our experiment. Similarly, if we watched their experiment we would note they were using unsynchronized (to us) clocks. But neither of us can convince the other of this flaw.

Suggested practice problems to ponder in Appendix A: P-2(a), (e); P-3(a), (b); P-16(c), (d); P-18(c), (d).

CHAPTER 4

Moving Objects Shrink

An argument was presented in Chapter 3 for the conclusion that the two reference frames agree on separations measured *perpendicular* to the direction of the relative motion. Now we examine the question of agreement on separations measured *parallel* to the direction of relative motion.

Let us make two marks on the railroad tracks separated by a distance L parallel to the tracks. We ask the *train* observers to determine the distance between these marks from their viewpoint. We are asking them to measure what to them is a *moving* separation. The two marks are at rest in our frame, so L represents a *rest* separation to us.

According to the train observers, the two marks are traveling toward the left at speed V. Suppose one of their clocks reads zero as the first mark passes this clock, and suppose the same clock reads t^* when the second mark passes it. Then they will say a time t^* was required for the space interval between the marks to pass their clock. The distance between the marks must be $L^* = Vt^*$, according to them.

Let us set up this situation, from our viewpoint, at the instant the left mark is under their clock A^*, as shown in Figure 4-1. *We* know it will take a time L/V for the right mark to be under the clock A^*. But since the A^* clock runs slow, this means it will read $(L/V)\sqrt{1-V^2/c^2}$ when the right mark is opposite it. The time interval t^* registered on their clock will be

$$t^* = \frac{L}{V}\sqrt{1-V^2/c^2}$$

Therefore, the train observers compute

$$L^* = Vt^* = V\frac{L}{V}\sqrt{1-V^2/c^2}$$

or

$$L^* = L\sqrt{1-V^2/c^2}$$

Putting the last expression in the form of words:

$$\text{Moving separation} = (\text{Rest separation})\sqrt{1-V^2/c^2} \qquad (4\text{-}1)$$

Thus the moving length is a smaller number than the rest separation.

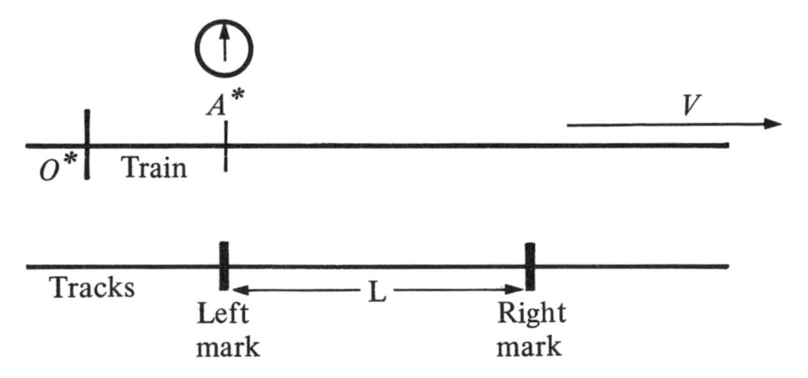

FIGURE 4-1:
Train Passing Over Marks on the Tracks (Our Viewpoint)

If we measure the distance between two marks painted on the train, to us the separation is a moving separation whereas the separation is a rest separation to the train observers. But equation (4-1) is still valid, and we will get a smaller value for the separation of these painted marks than the train observers do.

Suppose I have a meter stick in my hand. By definition, the stick's equilibrium length, determined by the forces holding the atoms in the stick together and by the way the stick was constructed, is 1 meter. So, of course, it is a meter long to us. But suppose I hand my meter stick to a train observer as he goes by. Now it is at rest with respect to him. The stick must have an equilibrium length of 1 meter from his viewpoint now that he has it, or else reciprocity would be violated. The two frames must agree on the *rest* length of a given stick or else the difference in *rest* lengths could be used as a criterion for giving preference to one reference frame over another.

The train observer must find the stick to be 1 meter long no matter in which direction he points it. Suppose we are determining its length *as it moves by in the train.* If it is pointed parallel to the direction of relative motion, we will measure it to be only $\sqrt{1-V^2/c^2}$ meters long. If it is pointed in a direction perpendicular to the relative motion, we will find it to be 1 meter long, because in Chapter 3 we showed that the frames must agree on separations along this direction.

We can determine the length of a stick pointed in an arbitrary direction. Suppose it is pointed in a direction illustrated in the top diagram of Figure 4-2, according to the train observer who is holding it. From our view, the component $L_{\parallel}$ gets shortened while the component $L_{\perp}$ does not. If we draw the moving rod, our diagram is as shown in the bottom diagram in Figure 4-2.

FIGURE 4-2:

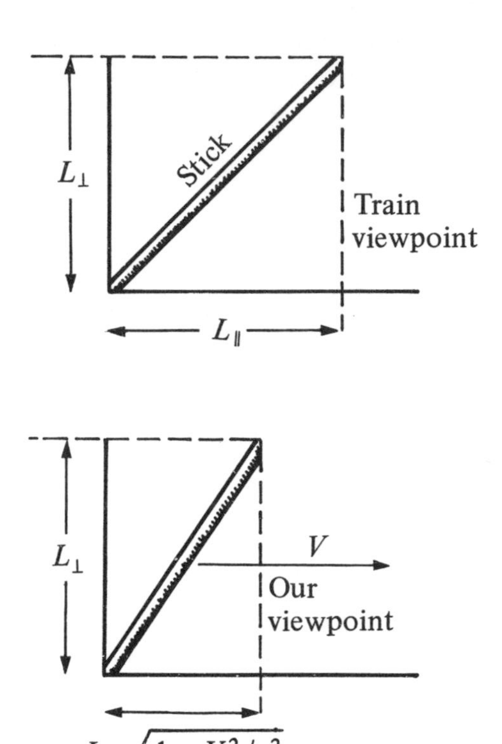

Lorentz Contraction

In summary, the dimension of a moving object measured parallel to the direction of relative motion is shortened compared with the corresponding dimension of the same equilibrium object at rest in our frame. This phenomenon is known as the *Lorentz-Fitzgerald contraction.* In this book it will be referred to simply as the Lorentz contraction.

Suggested practice problems to ponder in Appendix A: P-2(b), (c), (d), (f), (i); P-4(c); P-5; P-9(a); P-13(c); P-15(c).

CHAPTER 5

Another Look at Simultaneity

In Chapter 2 we were able to show qualitatively that two reference frames disagree on the simultaneity of two events and on the synchronization of sets of clocks in the two frames. Now we can obtain a quantitative expression for this disagreement. We derive the expression in two different ways.

Derivation 1

(based on comparing lengths in the two frames):

A rod with a rest length L_0 is placed along the tracks. We ask the train observers to measure its length as they go by. From Chapter 4 their answer must be $L^* = L_0\sqrt{1 - V^2/c^2}$. One way they can make this measurement is to locate the two ends of the rod (moving with respect to them) at some given time, say at $t^* = 0$. If they place two synchronized

> clocks a distance L^* apart (according to them), then the two ends of the rod should be under the clocks when both of them read $t^* = 0$.

Let us draw the situation from our viewpoint at the instant one end is under their left clock. In Figure 5-1 we have shown the two train clocks separated by a distance of $L_0\ (1 - V^2/c^2)$, according to us. The reason is that, according to them, their separation is $L^* = L_0\sqrt{1 - V^2/c^2}$. But because of the Lorentz contraction, their (moving) separation is only

$$L^*\sqrt{1 - V^2/c^2} = L_0(1 - V^2/c^2)$$

according to us. From the diagram we can see that the right train clock is a distance $L_0\ V^2/c^2$ from the right end of the rod. It will require a time interval equal to

$$\frac{\text{Distance}}{V} = \frac{L_0\ V^2/c^2}{V} = L_0\ V/c^2$$

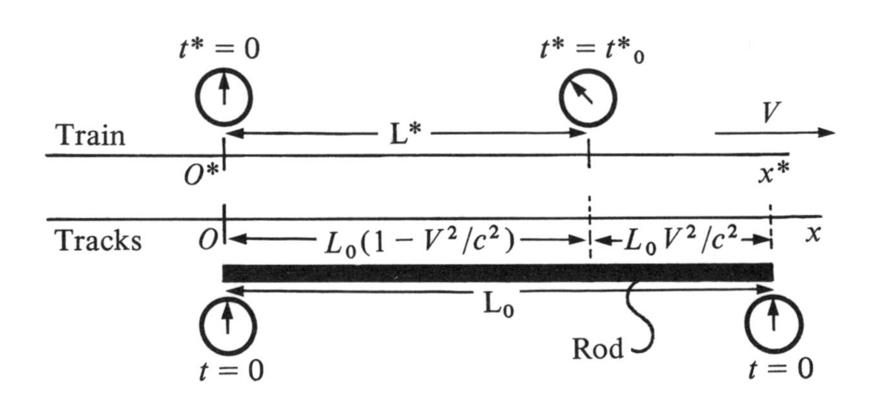

FIGURE 5-1:
Initial Situation from Viewpoint of Track Observers

for these two to get opposite each other. But because of the time dilation, the right train clock will tick off only a time interval equal to

$$(L_0 V/c^2)\sqrt{1-V^2/c^2} = L_0\sqrt{1-V^2/c^2}\,(V/c^2) = L^* V/c^2.$$

If the right train clock reads zero when it is opposite the right end of the rod, then it must read $t^*_0 = -L^* V/c^2$ at the instant shown in Figure 5-1.

Derivation 2

(based on measuring the speed of light in the two frames):

Consider two clocks on the train that are a distance L^* apart, according to the train. When the left clock on the train reads $t^* = 0$ it passes a mark on the tracks and a light beam from the mark is started toward the right clock. A time interval of L^*/c will elapse, according to the train observers, before the light beam gets to the right clock, so the right train clock must read L^*/c when the beam gets to it.

Let us look at the situation from our viewpoint. According to us the light beam has to catch up with the moving right clock. Figure 5-2 shows the situation, from our viewpoint, at the instant the light beam got started. Let t^*_0 be the reading on the moving right clock at the instant shown in the diagram. We can compute how long it will take for the light beam to catch up with this clock. The beam travels with speed c and the clock is traveling with speed V. So the beam is *gaining* on the clock at the relative speed of $c-V$. The clock was a distance $L^*\sqrt{1-V^2/c^2}$ ahead of the beam initially. Thus a time interval equal to (Distance to be gained)/(Relative speed) is required. This is

$\frac{L^* \sqrt{1-V^2/c^2}}{c-V}$. But the moving clock, because of time dilation, only ticks off $\sqrt{1-V^2/c^2}$ times this amount, or

$$\frac{L^* \sqrt{1-V^2/c^2}\ \sqrt{1-V^2/c^2}}{c-V}.$$

This equals

$$\frac{L^*(1-V^2/c^2)}{c-V}$$

which also equals

$$\frac{L^*}{c}\ \frac{(1-V^2/c^2)}{(1-V/c)}.$$

Consequently, if the right clock read t^*_0 to start with, its reading when the light beam catches it must be

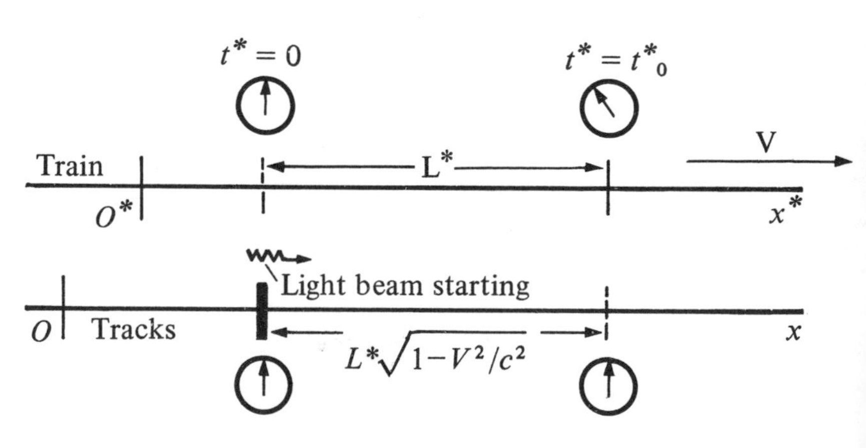

FIGURE 5-2:
Initial Situation from Viewpoint of Track Observers

$$t^*_0 + \frac{L^*(1 - V^2/c^2)}{c(1 - V/c)}$$

However, we already know that the final reading on this clock must be L^*/c in order to be consistent with the train observers' view of the situation. Therefore, the following equality must be valid:

$$t^*_0 + \frac{L^*}{c} \frac{(1 - V^2/c^2)}{(1 - V/c)} = \frac{L^*}{c}.$$

Subtract $\frac{L^*}{c} \frac{(1 - V^2/c^2)}{(1 - V/c)}$ from both sides. We get

$$t^*_0 = \frac{L^*}{c} - \frac{L^*}{c} \frac{(1 - V^2/c^2)}{(1 - V/c)} = \frac{L^*}{c} \left[1 - \frac{(1 - V^2/c^2)}{(1 - V/c)} \right]$$

Put the bracketed term in the right over a common denominator.

$$t^*_0 = \frac{L^*}{c} \left[\frac{1 - V/c - (1 - V^2/c^2)}{1 - V/c} \right]$$

$$= \frac{L^*}{c} \left[\frac{1 - V/c - 1 + V^2/c^2}{1 - V/c} \right]$$

or

$$t^*_0 = \frac{L^*}{c} \left[\frac{V^2/c^2 - V/c}{1 - V/c} \right] = \frac{L^* V (V/c - 1)}{c \cdot c\ (1 - V/c)} = -\frac{L^* V}{c^2}$$

so $t^*_0 = -\frac{L^* V}{c^2}$.

This is exactly the same result yielded by Derivation 1. According to us, the train clocks further forward in the direction of relative motion (to the right in this situation) read *earlier* than the clocks behind. In reversing the roles of the two frames, exactly the same argument forces the train observers to conclude that our clocks further forward in the direction of relative motion (toward the left from their view) read earlier than our clocks behind. Again, reciprocity is satisfied.

Suggested practice problems to ponder in Appendix A: P-9; P-13(b); P-15(b); P-17(c).

CHAPTER 6

Comparing Speeds

Comparisons of speeds (the speed of light, in particular) in different reference frames have played a fundamental role in many of our considerations in the preceding chapters. Now we are in a position to obtain a general expression that relates the speed of an object measured in one frame to the speed of the same object from the viewpoint of another reference frame.

Again, let us think in terms of the track frame (our frame) and the train frame. Suppose an object is moving toward the right at speed v^* according to the train observers. Let v be the speed of the same object from our frame's viewpoint. V is the speed of the train relative to us.

For simplicity let us assume the object passes the train's origin clock when it reads $t^* = 0$. Then the object should pass the train's clock situated at $x^* = L^*$ when it reads L^*/v^* if the train observers are to conclude that the speed of the object is v^*. Figures 6-1 and 6-2 show the initial and final situations *from our view.* We have used the relation for disagreement on simultaneity (see Chapter 5) to assign an initial value of $t^* = -L^* V/c^2$ to the $x^* = L^*$ clock.

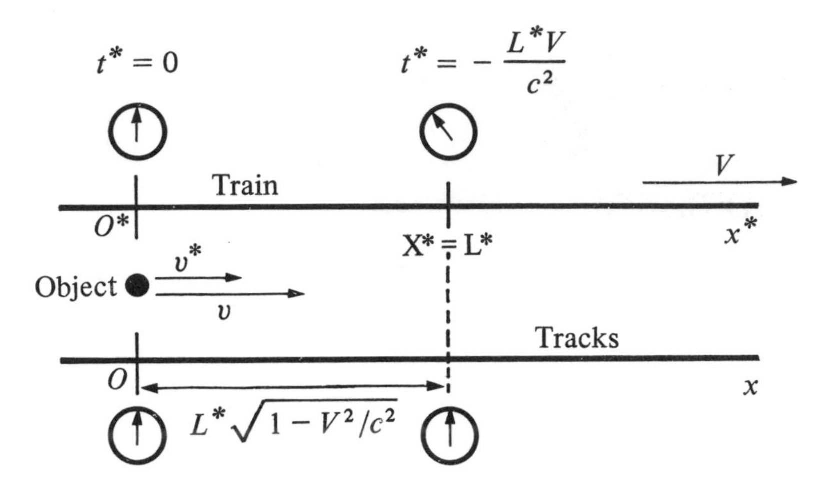

FIGURE 6-1:
Initial Situation: Object Passing the $x^* = 0$ Clock. (Our Viewpoint)

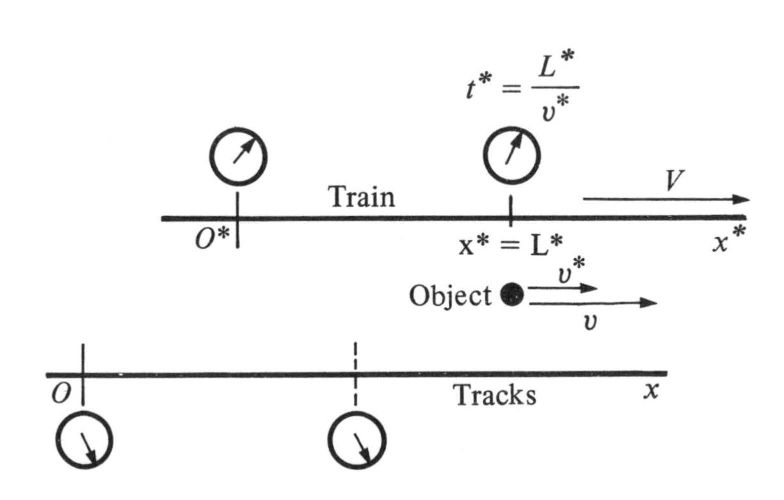

FIGURE 6-2:
Final Situation: Object Passing the $x^* = L^*$ Clock. (Our Viewpoint)

Between the initial and final situations, the $x^* = L^*$ clock must tick off $\dfrac{L^*}{v^*} + \dfrac{L^* V}{c^2}$ seconds in order to be reading L^*/v^* in the final situation. But it is running slow, according to us, so the time interval between the initial and final situations is

$$\Delta t = \frac{L^*/v^* + L^* V/c^2}{\sqrt{1 - V^2/c^2}} = \frac{L^*}{\sqrt{1 - V^2/c^2}} \left(\frac{1}{v^*} + \frac{V}{c^2}\right)$$

Therefore, the distance traveled by the object between the two situations is

$$\text{Distance between the initial and final positions of the object} = v\Delta t = \frac{vL^*}{\sqrt{1 - V^2/c^2}} \left(\frac{1}{v^*} + \frac{V}{c^2}\right) \tag{6-1}$$

Relativistic Addition Law of Velocities

The object has to catch up with the $x^* = L^*$ clock, so we can get another expression for the distance traveled by the object between the initial and final situations by computing how far the $x^* = L^*$ clock is from our origin in the final situation. It started out at a distance $L^* \sqrt{1 - V^2/c^2}$ away from our origin. It is traveling at speed V, so it would travel an extra distance of

$$V\Delta t = \frac{VL^*}{\sqrt{1 - V^2/c^2}} \left(\frac{1}{v^*} + \frac{V}{c^2}\right)$$

Therefore, the total distance from our origin to the $x^* = L^*$ clock at the end is

Distance between the initial and final positions of the object =

$$= L^* \sqrt{1 - V^2/c^2} + \frac{VL^*}{\sqrt{1 - V^2/c^2}} \left(\frac{1}{v^*} + \frac{V}{c^2} \right) \qquad (6\text{-}2)$$

Now equate the two equivalent expressions in equations (6-1) and (6-2).

$$\frac{vL^* \left(\frac{1}{v^*} + \frac{V}{c^2} \right)}{\sqrt{1 - V^2/c^2}} = L^* \sqrt{1 - V^2/c^2} + \frac{VL^*}{\sqrt{1 - V^2/c^2}} \left(\frac{1}{v^*} + \frac{V}{c^2} \right)$$

Multiply both sides by

$$\frac{\sqrt{1 - V^2/c^2}}{L^* \left(\frac{1}{v^*} + V/c^2 \right)}$$

The L^* cancels out.

We get

$$v = \frac{1 - V^2/c^2}{\frac{1}{v^*} + V/c^2} + V$$

Put the right side over a common denominator:

$$v = \frac{1 - V^2/c^2 + \frac{V}{v^*} + V^2/c^2}{\frac{1}{v^*} + V/c^2} =$$

$$\frac{1 + \dfrac{V}{v^*}}{\dfrac{1}{v^*} + V/c^2} =$$

$$\frac{\dfrac{v^* + V}{v^*}}{\dfrac{1 + Vv^*/c^2}{v^*}} =$$

$$\frac{v^* + V}{1 + Vv^*/c^2}$$

$$\text{So} \quad v = \frac{v^* + V}{1 + \dfrac{v^* V}{c^2}} \tag{6-3}$$

This is known as the relativistic addition law of velocities.

If the motion of the object is to the left in either frame, then the corresponding v or v^* should be used with a negative sign.

Equation (6-3) can be solved for v^*. The result is

$$v^* = \frac{v - V}{1 - \dfrac{vV}{c^2}} \tag{6-4}$$

Both (6-3) and (6-4) are valid only when v^* and v are parallel (or antiparallel) to the direction of the relative motion of the two frames (i.e., parallel to the relative velocity V). A much more complicated relation can be derived that is valid even when v or v^* is not parallel to V. However this more general relation will not be considered here; the interested reader is referred to one of the more comprehensive texts on relativity.

Apparent Violation of Reciprocity

A comparison of equation (6-3) with equation (6-4) discloses an apparent violation of reciprocity. One of the equations has plus signs and the other has minus signs. Why couldn't the two frames argue about whether they preferred plus or minus signs in their formula, and hence have a criterion for a preferred frame?

The lack of reciprocity comes about because the same quantity V is used in both formulas. V has the significance of the velocity of the train relative to the tracks, and has a direction toward the right. V is not the relative *velocity* of the tracks relative to the train, because the latter relative velocity has a direction toward the left. Let V_{rel} equal the velocity of the train relative to the tracks. Then $V_{rel} = V$. Let V^*_{rel} equal the velocity of the tracks relative to the train. Then $V^*_{rel} = -V$. In terms of these new quantities, (6-3) and (6-4) become

$$v = \frac{v^* - V^*_{rel}}{1 - \dfrac{v^* V^*_{rel}}{c^2}} \qquad (6\text{-}5)$$

$$v^* = \frac{v - V_{rel}}{1 - \dfrac{v V_{rel}}{c^2}} \qquad (6\text{-}6)$$

Now we see that reciprocity is restored in terms of the corresponding relative velocities.

Old Form of the Addition Law of Velocities

According to the *old* ideas, it was assumed that the correct addition law of velocities was merely

$$v = v^* + V \qquad (6\text{-}7)$$

This was what we were using implicitly in Chapter 1 in our discussion of a person walking in a train.

From equation (6-3) we can see that if $v^* \ll c$ and $V \ll c$ (which is certainly true for most everyday situations), then the term $v^* V/c^2$ in the denominator would be very small compared with unity. Under these circumstances (6-3) becomes very close to (6-7), and would account for why (6-7) is very accurate for most all everyday situations. But for speeds near the speed of light, (6-3) may be very different from (6-7).

Verifying That if $v^* = c$, then $v = c$

Suppose the object is traveling at the speed of light c according to the train; that is, $v^* = c$. Then it should also be traveling at speed c according to us. Let us check this out. The question is: Does (6-3) imply that if $v^* = c$, v must also equal c? We put $v^* = c$ into

$$v = \frac{c + V}{1 + cV/c^2} + \frac{c + V}{1 + V/c} = \frac{c + V}{\dfrac{(c + V)}{c}} = c$$

This confirms our expectation.

Suggested practice problems to ponder in Appendix A: P-2(g), (h); P-4(a), (b); P-10; P-16(f); P-17(e).

CHAPTER 7

Derivation of the Lorentz Transformation

Most of the discussion in Chapters 2 through 6 involved comparisons of situations and sequences of events from the viewpoint of two different reference frames. Each reference frame could set up a Cartesian coordinate system [x, y, z-axes in our (unstarred) frame; x^*, y^*, z^*-axes in the train (starred) frame] with respect to which events are located in space, and a collection of synchronized clocks to determine the time of the events. Each event would then be assigned four numbers x, y, z, t, which tell where and when the event occurred from the unstarred frame's viewpoint; the *same* event would also be assigned four numbers x^*, y^*, z^*, t^*, which tell where and when the event occurred from the starred frame's viewpoint. If we could give a prescription for converting the x, y, z, t numbers into the correct x^*, y^*, z^*, t^* numbers, then we could use the prescription to transform a space-time description from the unstarred view into a space-time description of the same events from the starred view.

We could have derived the correct transformation rule directly from the principle of relativity and its corollaries. Then the transformation equations could have been used to deduce the time dilation, the Lorentz contraction, and the disagreement on simultaneity. In fact, this is the way relativity theory is developed in a majority of relativity texts, and there is no doubt about the conciseness and elegance of this approach. The reader is referred to one of the standard texts for more details.

The author has found merit in the alternate, more long-winded approach presented in Chapters 2-6. Here, the consequences of relativity are developed one at a time in such a way as to show explicitly how they are interrelated. In this way the reader can develop insight into these strange concepts in a gradual step-by-step manner. The elegance and conciseness of the more usual approach tend to disguise some of these features, largely because the great power of mathematics is used to obtain all the results in one rather abstract grand derivation. Even after mastering the derivation, the student of relativity discovers that he must devote much additional time to digesting the significance of the transformation equations.

Now we can derive the transformation equations on the basis of the results we established previously. Again, for simplicity we suppose the two origin clocks happen to read zero when they pass each other. In addition, we choose the x- and x^*-axes along the same straight line and parallel to the direction of the relative velocity. The other corresponding axes are also mutually parallel. Figure 7-1 shows the situation from our (unstarred) view at the instant some event is happening at some x, y, z, t. Let x^*, y^*, z^*, t^* be the starred readings for this event. Let t^*_{0*} equal the reading on the starred origin's clock when the event occurred (from our view). Also, let x_{0*} be the *x-reading* of the starred origin's clock at this time. The illustration shows the x^*-

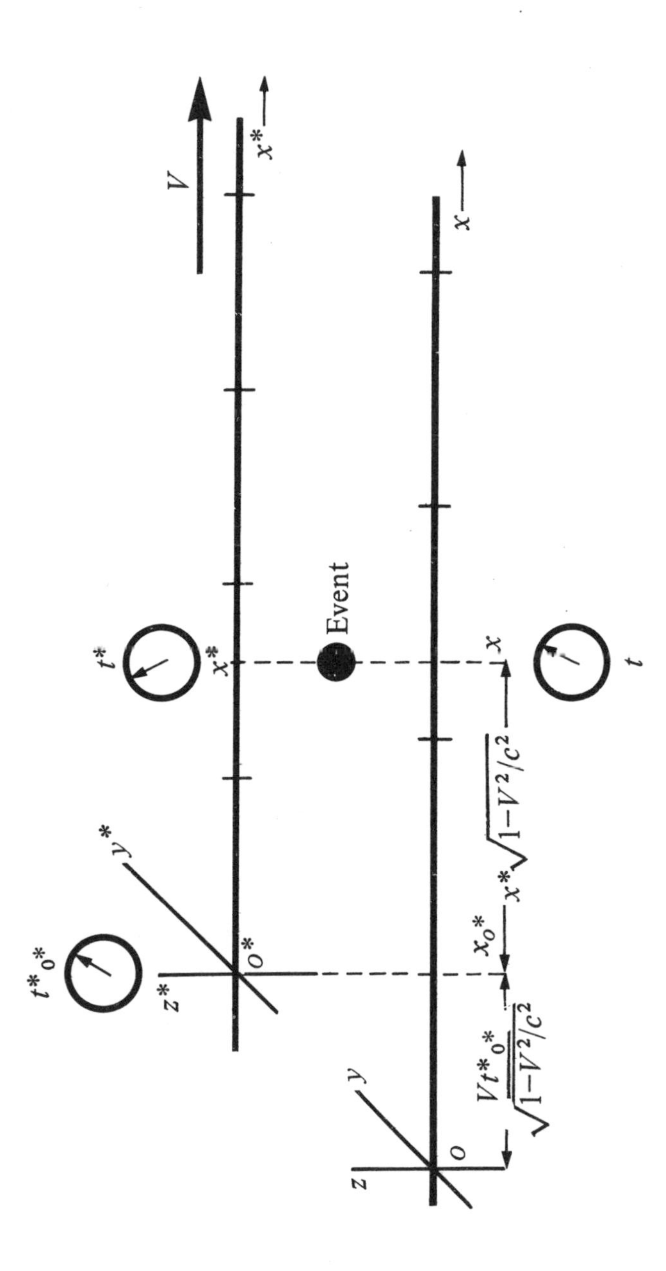

FIGURE 7–1:
Situation when an Event Occurs (Our Viewpoint)

and x-axes as slightly displaced (to avoid confusion in the diagram) but it is to be understood that they are collinear.

First of all, we already know there is agreement on distance measured in the y and z directions (see Chapters 3 and 4). Consequently, $y = y^*$, $z = z^*$ are two of the equations we seek.

From the lack-of-synchronization formula obtained in Chapter 5 (remember, x^* is the starred distance between O^* and the event at x^*, as measured along the x^*-axis),

$$t^* = t^*_{o^*} - x^* V/c^2$$

or (7-1)

$$t^*_{o^*} = t^* + x^* V/c^2$$

From the Lorentz contraction formula, if the starred distance between the event and O^* is x^*, then the unstarred distance between these is $x^* \sqrt{1 - V^2/c^2}$. The total distance from our origin O to where the event occurs is (see Figure 7-1)

$$x = x_{o^*} + x^* \sqrt{1 - V^2/c^2} \tag{7-2}$$

But the O^* clock was at our origin O at $t = 0$ and at time t it will be at a distance from our origin of

$$x_{o^*} = Vt \tag{7-3}$$

At the instant shown in Figure 7-1, the O^* clock reads $t^*_{o^*}$. But it runs slow, according to us, so the time t must satisfy

$$t = \frac{t^*_{o^*}}{\sqrt{1 - V^2/c^2}} \tag{7-4}$$

Substitute this into (7-3). We get

$$x_{o^*} = \frac{Vt^*_{o^*}}{\sqrt{1 - V^2/c^2}}$$

Now put this into (7-2).

$$x = \frac{Vt^*_{o^*}}{\sqrt{1-V^2/c^2}} + x^*\sqrt{1-V^2/c^2} \tag{7-5}$$

From (7-1) we can replace $t^*_{o^*}$ with $t^* + x^*V/c^2$ in (7-5):

$$x = \frac{V(t^* + x^*V/c^2)}{\sqrt{1-V^2/c^2}} + x^*\sqrt{1-V^2/c^2}$$

Put the right side over a common denominator:

$$x = \frac{Vt^* + x^*V^2/c^2 + x^*(1-V^2/c^2)}{\sqrt{1-V^2/c^2}} =$$

$$\frac{Vt^* + x^*V^2/c^2 + x^* - x^*V^2/c^2}{\sqrt{1-V^2/c^2}}$$

or

$$x = \frac{x^* + Vt^*}{\sqrt{1-V^2/c^2}}$$

This is another of the equations we were seeking.

Now from (7-4) we have

$$t^*_{o^*} = t\sqrt{1-V^2/c^2}$$

Put this into (7-1):

$$t\sqrt{1-V^2/c^2} = t^* + x^*V/c^2$$

or

$$t = \frac{t^* + x^*V/c^2}{\sqrt{1-V^2/c^2}}$$

The Lorentz Transformation Equations

Collecting our results, we have

$$
\begin{aligned}
x &= \frac{x^* + Vt^*}{\sqrt{1 - V^2/c^2}} \\
y &= y^* \\
z &= z^* \\
t &= \frac{t^* + Vx^*/c^2}{\sqrt{1 - V^2/c^2}}
\end{aligned}
\qquad (7\text{-}6)
$$

Equations (7-6) are known as the *Lorentz transformation equations.* They relate the x^*, y^*, z^*, t^* numbers to the corresponding x, y, z, t numbers for any event.

Another Form of the Lorentz Transformation Equations

It is an interesting problem in algebra to solve the foregoing equations for the starred quantities in terms of the unstarred. The result is

$$
\begin{aligned}
x^* &= \frac{x - Vt}{\sqrt{1 - V^2/c^2}} \\
y^* &= y \\
z^* &= z \\
t^* &= \frac{t - Vx/c^2}{\sqrt{1 - V^2/c^2}}
\end{aligned}
\qquad (7\text{-}7)
$$

Equations (7-7) are exactly equivalent to (7-6), except that they are written in a different form. Note that the form of (7-7) differs from (7-6) in that $(-V)$ replaces $(+V)$. One might think this to be a violation of the reciprocity feature required by the principle of relativity. But if we introduce $V^*_{\text{rel}} = -V$ as discussed in Chapter 6, we will find the form of the equations to be compatible with reciprocity.

Galilean Transformation Equations

According to prerelativity concepts, it had been thought that the following transformation equations were valid:

$$x = x^* + Vt^*$$
$$y = y^*$$
$$z = z^*$$
$$t = t^*$$

or

$$x^* = x - Vt$$
$$y^* = y$$
$$z^* = z$$
$$t^* = t$$

These are called the *Galilean transformation equations.*

Under conditions where $V \ll c$ (which means $\sqrt{1 - V^2/c^2}$ is very close to unity) and Vx^*/c^2 is very small, the Lorentz transformation equations are almost the same as the Galilean transformation equations. For ordinary everyday situations, the Galilean transformation equations would be very, very accurate. This explains why no one had noticed anything wrong with them for such a long time.

As mentioned before, many elementary relativity texts show how to use the Lorentz transformation equations to deduce the properties described in Chapters 2–6. We have done things in reverse order, and showed how the Lorentz transformation equations can be deduced from the results established in the preceding chapters.

SECTION II

DETAILED EXAMPLES: QUESTIONS AND ANSWERS

CHAPTER 8

A High-Speed Trip to a Star (or How to Live a Long Time)

With this chapter we begin a detailed discussion of some informative examples of the application of relativistic concepts in a variety of situations. A careful study of these examples should deepen your understanding of these concepts.

For the first example, we consider a rocket trip from the earth to a star 8 light-years from the earth. (A light-year is the distance light travels in one year.) The rocket travels at a speed of $4c/5$; hence the square root factor $\sqrt{1-V^2/c^2}$ equals $3/5$. Also, according to us on the earth, 10 years should be required for the rocket to go from the earth to the star.

Figure 8-1 shows the initial situation, from our viewpoint, as the rocket starts out toward the star. We assume the earth's clock and the rocket clock both read zero at the start. A clock is shown at the star, which is synchronized with the earth's clock.

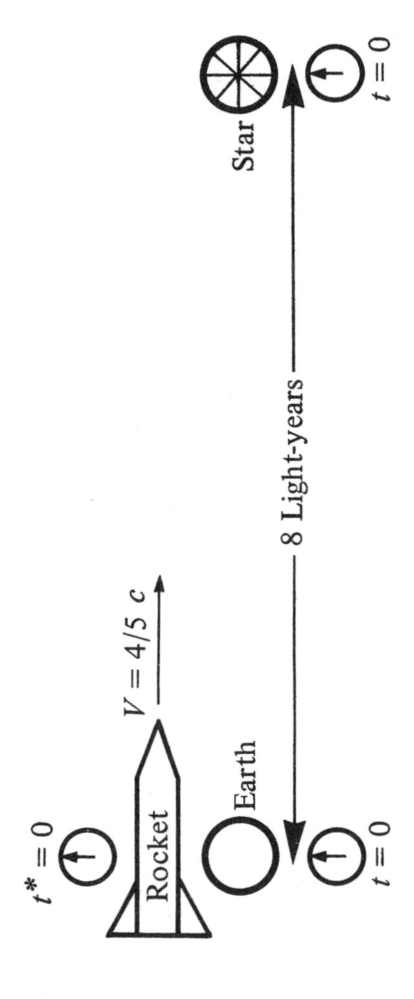

FIGURE 8-1:
Initial Situation from Earth's Viewpoint

Earth Observers' Analysis

Question 8-1:

What will the rocket clock read when it gets to the star?

According to us, it will take 10 years for the rocket to get to the star. But the rocket clock runs slow (according to us) by a factor of 3/5. Hence the rocket clock reading will be $10 \times 3/5 = 6$ years. Thus the answer to Question 8-1 is 6 years.

Figure 8-2 represents the situation when the rocket is at the star, from our viewpoint.

Rocket Observers' Analysis

Question 8-2:

Let us look at the situation from the viewpoint of an inertial reference frame moving at a speed of $4c/5$ to the right with respect to the earth. Thus the rocket is at rest in this inertial frame during its journey to the star. How does this inertial frame account for the fact the rocket clock registered only 6 years when it arrived at the star?

Let us describe the entire situation from the viewpoint of the rocket inertial frame. According to it, the rocket is at rest and both the earth and the star are traveling to the left at a speed of $4c/5$. The *rest* separation between the earth and the star measured by the rocket frame is a *moving* separation and undergoes the Lorentz contraction. Thus the distance between the earth and the star from the viewpoint of the rocket's frame is $8 \times 3/5 = 24/5$ light years. This is shown in Figure 8-3.

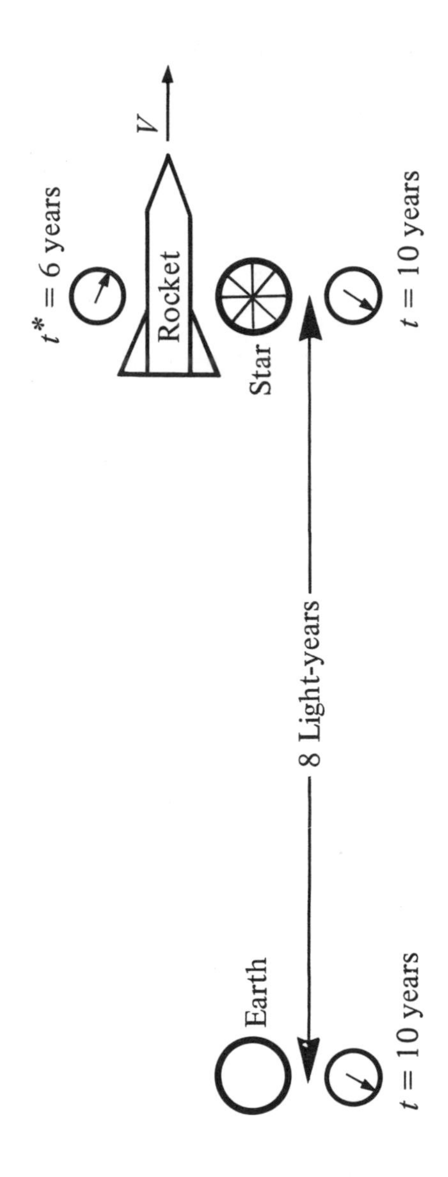

FIGURE 8–2:
Rocket at the Star (Earth's Viewpoint)

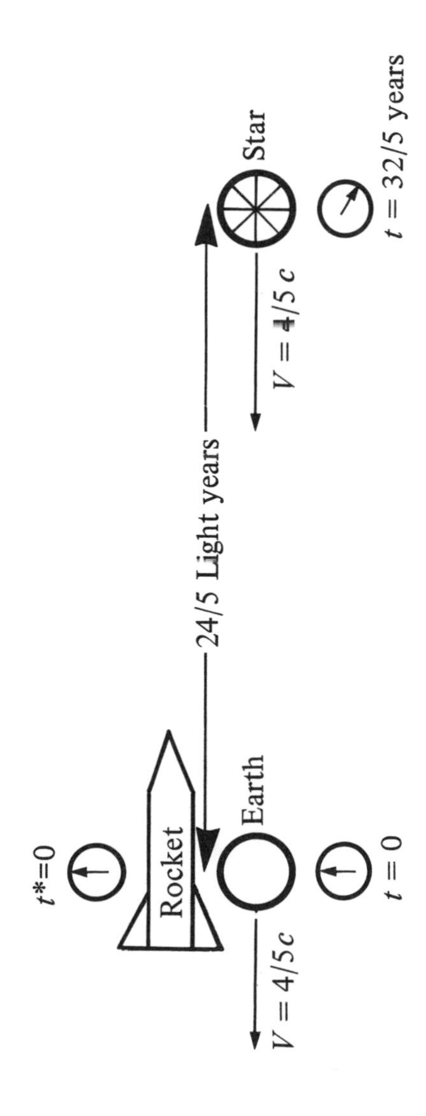

FIGURE 8–3:
Initial Situation From the Viewpoint of the Rocket Frame

From the rocket observers' viewpoint, it should take only 6 years for the star to travel the 24/5 light-years at a speed of $4c/5$. Therefore, 6 years after the initial situation the star should be at the same location as the rocket.

Comparison of the Two Explanations

The explanations of the earth observers and the rocket observers are much different. They do not argue about the readings on the clocks when they are opposite each other; in fact, there must be no argument on these objective data. The difference is only in the interpretation or explanation of the data. The earth observers say the rocket clock ticked off only 6 years because it was running slow and therefore registered only 6 years rather than the 10 years actually involved. The rocket observers say their clock was running correctly (in fact, they claim the earth's clock was running slow) but that the distance involved was only 24/5 light-years rather than 8 light-years, so only 6 years were required to traverse this shorter distance. The earth observers use time dilation in their explanation, whereas the rocket observers use the Lorentz contraction.

In Figure 8-3, we have shown the clock at the star to read 6.4 years at the instant shown ($t^* = 0$) from the rocket observers' view. This arises from the fact that the clocks synchronized according to the earth's view are out of synchronization according to the rocket's view. The disagreement in synchronization is VL/c^2 (see Chapter 5), which equals $4/5 \times 8 = 32/5$ years. If we were to ask the rocket observers to compute what the star clock will read when it is opposite the rocket, they would make their computation as follows:

Six years will elapse. However, the star clock runs slow, and therefore would tick off $6 \times 3/5 = 18/5$ years. But if it reads $32/5$ years at the start, its final reading will be $32/5 + 18/5 = 50/5 = 10$ years.

Both frames of reference can explain very nicely that the rocket clock would read 6 years and the star clock would read 10 years when they are opposite each other. Each, however, insists that the other frame's clocks are running slow.

Disagreement on Simultaneity

Question 8-3:

At the instant the rocket reaches the star, what does the clock back on earth read?

Different inertial frames disagree on simultaneity, so the answer depends on who is giving the answer. The earth observers say the answer is 10 years, as discussed earlier. The rocket inertial frame observers claim 6 years have elapsed, but that the earth's clock was running slow and consequently reads only $3/5 \times 6 = 3.6$ years. Thus there is a wide discrepancy between the two frames as to what event back on the earth is simultaneous with the rocket reaching the star.

Round-Trip Journey

Suppose the rocket quickly turns around after it reaches the star and, keeping its speed of $4c/5$ relative to the earth, it returns to the earth.

Question 8-4:

What will the rocket and the earth clocks read when they get back together again?

This type of question has been the source of considerable controversy. Because of this, the answer will be obtained by five different methods in order to demonstrate that there should be no dispute over the correct answer.

Analysis of Journey From Earth's Viewpoint

Method 1:

Analysis of the journey from the viewpoint of the earth's inertial reference frame

We describe the entire round trip from the viewpoint of the earth's observers, who constitute an inertial reference frame. The round trip requires 20 years, so the earth's clock should read 20 years at the end of the journey. But the rocket clock was running slow by a factor of 3/5, so the rocket clock should read $20 \times 3/5 =$ 12 years upon its return. All this is a straightforward application of the time dilation effect.

However, we know that the rocket observers on the rocket's outward journey insist that the earth's clock was running slow. Also, on the return journey the rocket observers presumably would still maintain that the earth's clock was running slow. Therefore, it might seem reasonable for the rocket observers to expect to find the earth's clock reading less than the rocket clock when they get back together. But the previous argument based on the earth observers' viewpoint gives the opposite result. One or the other of the arguments must be wrong because the actual readings on the clocks could be observed when they are back together, and *everyone* would have to accept the validity of the observed readings.

Twin Paradox

There seems to be a paradox here; that is, one argument says the earth's clock should read more than the returned rocket clock and the other argument says the reverse is true. The observers on the earth should age in the usual way relative to the earth's clock, whereas the rocket observers should age in a similar fashion relative to the rocket clock. According to the earth's observers, if one member of a pair of twins embarked on the rocket journey, the one who stayed behind should age 20 years during the journey while the one taking the rocket trip would age only 12 years. However, according to the other argument, the earth twin would be younger than the rocket twin. This quandary is known as the "twin paradox."

Argument Over the Twin Paradox

During the years since 1905 some students of relativity have not accepted either argument, and have maintained that the two clocks should read exactly the same when they get back together (and therefore the two twins would have aged by exactly the same amount). However, no acceptable way of fitting this conclusion into the rest of relativity theory has been found, and in recent years there has been experimental verification of the correctness of the result implied by the earth observers' explanation (see Chapter 14). There seems to be no real doubt that the correct answer for the readings is: earth's clock – 20 years; rocket clock – 12 years.

Behavior of Accelerated Clocks

We have assumed that nothing particularly strange happened to the rocket clock during the short time it was accelerating while it turned around at the star. Some have argued that special relativity theory cannot treat the behavior of *accelerated* clocks. However, this is not true. It is true that not all clocks would register the same amount of time if they were accelerated together, even though they did run at the same rate when moving at the same *constant* velocity. For example, the light clocks used in Chapter 3 could be constructed with various values of the mirror separation L. All of them could be calibrated to register the correct time in their inertial frame. But if these clocks were accelerated, they would register different time intervals. If we want our light clocks to exhibit the time dilation effect in such a way that the time dilation factor depends only on the instantaneous speed and not on the acceleration, then we must use light clocks with very small L. But even for large L, one could compute the effect of acceleration on the rate of running of the clock. In any case there is no physical mechanism for *any* clock to undergo a significant change in its reading during the short turnaround time (short compared with the 20 years required for the journey). Thus no important error is made in neglecting the turnaround effects on the rocket clock; in fact, *no* error would be made if our clock were a light clock with infinitesimal mirror spacing.

In a sense the biological aging process is one form of (crude) clock. The aging process rate might be somewhat out of phase with the infinitesimal light clock during the acceleration period at the star, but assuming the turnaround to take place in a very short time interval compared with 20 years, any discrepancy between the rocket clock time interval and the rocket observers' biological aging time would be negligible. We assume, of course, that the acceleration is not sufficient to kill the biological clock!

There seems to be no doubt, then, that the twin taking the journey would arrive back at the earth only 12 years older, even though 20 years had elapsed on the earth. This offers the intriguing possibility of building an even faster rocket; for example, so that $\sqrt{1-V^2/c^2} = 1/1000$. Then a round-trip journey requiring 1000 years according to the earth's observers would age the rocket observer only 1 year. This would enable a person to be alive to see how things are on earth 1000 years from now, a feat not possible for earthbound people now living.

A Practical Difficulty

Before getting too excited about this possibility, however, one should be aware of a major practical difficulty in getting a rocket up to these very high speeds. There is about one hydrogen atom per cubic centimeter in the space between the stars in the galaxies. This represents a fantastically good vacuum (there are billions of atoms per cubic centimeter in the best vacuum we can produce on earth) but a high-speed rocket would have to push through this matter. Effectively, the rocket would be bombarded by very-high-speed protons and electrons. It seems to be an insurmountable problem to shield the rocket occupants from the lethal radiation produced by this bombardment.

Beware of Non-Inertial Reference Frames!

What was wrong with the explanation proposed by the rocket observers during the round-trip journey? Why don't they find the earth's clock to read less than theirs when they get back, if there really is a time dilation? The source of the difficulty is that the rocket observers are *not* inertial observers during the *entire* journey. In fact, they were in one inertial frame going to the star and in another inertial frame on the return trip. If they analyze what is happening during part of the journey from the viewpoint of one inertial frame and analyze the rest of the journey from the viewpoint of another reference frame, then they are apt to find their overall deductions to be incorrect. Their deductions during the first part of the journey are not compatible with the viewpoint of the second inertial frame; attempting to fit them together might lead to paradoxical conclusions. And this is exactly why they find their expectations, based purely on time dilation effects and ignoring any problem with changing inertial frames during the journey, are not fulfilled.

However, if the entire process is described from any *one* inertial frame, then the result for what the clocks should read after the journey will be independent of which inertial frame is used. This will be illustrated in the next two methods for answering Question 8-4.

Analysis of Journey from the Viewpoint of a Second Inertial Reference Frame

Method 2:

Analysis of the journey from the viewpoint of the rocket's initial inertial reference frame

Choose an inertial frame traveling to the right with speed $4c/5$ relative to the earth. The rocket is at rest in this frame on its journey *to* the star. The earth and the star are separated by a distance of 4.8 light-years and are moving to the left at a speed of $4c/5$, from the viewpoint of this reference frame. Let us refer to this frame as the starred reference frame. See Figure 8-3 for a diagram of the initial situation.

The rocket clock will remain at rest for the 6 years it takes to get opposite the star. But then the rocket suddenly accelerates to the left and begins to gain on the earth. Figure 8-4 shows the situation at the instant $t^* = 6$ years, when the rocket starts on its return journey to the earth.

Let us compute the speed of the rocket relative to the starred reference frame during the return journey of the rocket. We know that it travels with a velocity of $-4c/5$ relative to the earth (the minus sign indicates it is traveling toward the left). The relative speed between the two frames is $V = 4c/5$. Substitute into the addition law of velocities, equation (6-4):

$$v^* = \frac{v - V}{1 - \dfrac{vV}{c^2}} = \frac{-\dfrac{4}{5}c - \dfrac{4}{5}c}{1 + \dfrac{4}{5} \cdot \dfrac{4}{5}} = \frac{-\dfrac{8}{5}c}{\dfrac{41}{25}} = -\frac{40}{41}c$$

Thus, on the return journey the rocket is traveling to the left with a speed of $\frac{40}{41}c$ relative to the starred inertial frame.

How long does it take for the rocket to catch up with the earth (according to the starred frame)? It must gain a distance of 24/5 light-years and it is going $(40/41 - 4/5)c = (36/205)\,c$ faster than the earth. Therefore, $24/5 \div 36/205 = 82/3$ years are required to catch up. The earth's clock is running slow by a factor of 3/5, so it will tick off $3/5 \times 82/3 = 82/5$ years. But it read 18/5 years when the rocket started back (see the answer to Question 8-3). So the earth's clock should read $18/5 + 82/5 = \dfrac{100}{5} = 20$ years when the rocket gets back to the earth.

The rocket clock on its return journey runs slow by a factor of $\sqrt{1 - (40/41)^2}$, according to the starred inertial frame. But $\sqrt{1 - (40/41)^2} = 9/41$, and the rocket clock ticks off $9/41 \times 82/3 = 6$ years during the 82/3 years it

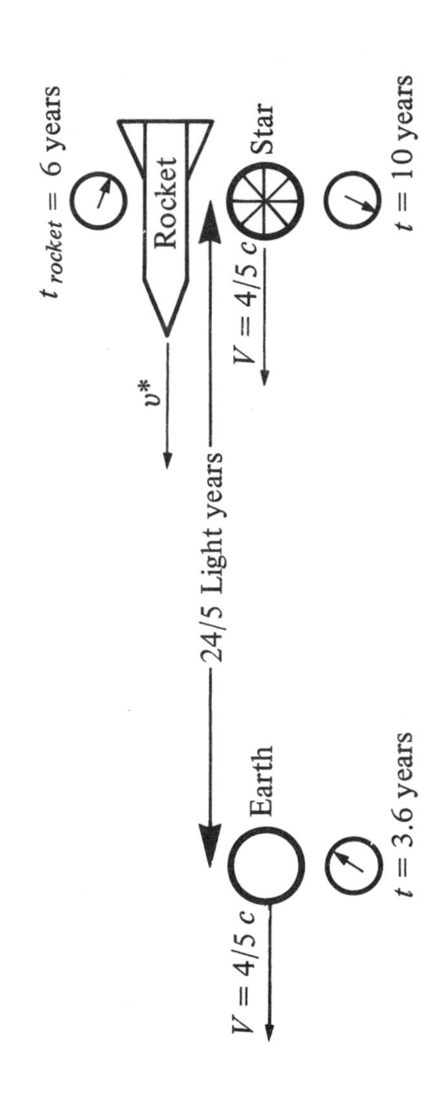

FIGURE 8–4:
Rocket Starting its Return Trip (Viewpoint of Starred Inertial Frame)

took to catch up to the earth. However, the rocket clock read 6 years when it started its return journey, and thus should read $6+6=12$ years when it gets back. According to the starred inertial frame, the earth's clock reads 20 years and the rocket clock reads 12 years when they get back together. This is in agreement with the values obtained by Method 1.

Let us briefly summarize these last results. The entire journey took $33\frac{1}{3}$ years according to the starred frame. The earth's clock ran slow by a factor of 3/5, and consequently registered $3/5 \times 33\frac{1}{3} = 20$ years. The rocket clock was at rest in the starred frame for the first 6 years. Another $27\frac{1}{3}$ years were required for the rocket to catch the earth and the rocket was traveling at a speed of $(40/41)c$ during this time interval. It was running slow by a factor of 9/41 during these $27\frac{1}{3}$ years, and it therefore ticked off 6 more years. The final reading on the rocket clock was 12 years.

Analysis of Journey from the Viewpoint of a Third Inertial Reference Frame

Method 3:

Analysis of the journey from the viewpoint of the rocket's final inertial reference frame

Consider an inertial reference frame moving toward the left at a speed of $4c/5$ relative to the earth. Let us call this the primed reference frame. In this frame the rocket is at rest during the return portion of its round trip. The description of the journey from the primed frame's view is similar to the description given by the starred frame in Method 2, except that the features of the outgoing portion are interchanged with the ingoing portion.

From the primed frame's view, the earth and the star are traveling at $4c/5$ toward the right and are separated by a distance of 4.8 light-years. The rocket starts out at a speed of $(40/41)c$ and takes 82/3 years to catch the star. It ticks off only $9/41 \times 82/3 = 6$ years. Then it comes to rest in the primed frame. Six more years are needed for the earth to get to the rocket. So the rocket reads 12 years when the earth and rocket are together again. A total of $6 + 82/3 = 33\frac{1}{3}$ years elapsed in the process. But the earth's clock ran slow by a factor of 3/5, and so the earth's clock read $3/5 \times 33\frac{1}{3} = 20$ years at the end.

We have shown by Methods 1, 2, and 3 that all three inertial reference frames predict exactly the same result for the earth's and the rocket's clocks at the end of a journey. Every other inertial reference frame also would yield the same results.

Analysis of Journey from Data Obtained by an Earth Observer Looking through a Telescope

Method 4:

Analysis of the round trip in terms of the data obtained by an earth observer viewing the rocket's clock through a telescope

We assume that the rocket clock is continually observed through a telescope by an observer on earth. The earth observer keeps a record of what he sees through the telescope at various times by his own clock. Let t be the time on the earth observer's clock and let t^* be the time *seen* on the rocket clock by the earth observer at time t.

During the outward journey the time t^* will be less than

t for two reasons: the time dilation effect, and the increasing time lag involved for the light from the rocket to get back to the earth.

It takes 10 years for the rocket clock to get to the star, and it will take 8 more years before the image of the rocket at the star gets back to the earth. After a total of 18 years, the earth observer will *see* the rocket clock at the star reading 6 years. The rocket clock appears through the telescope to be running slow by a factor of 1/3 (ticks off 6 years in 18 years). Also, its apparent recession speed was $(8/18)c = (4/9)c$ (it appeared to travel 8 light-years in 18 years).

Let us check the foregoing results by a direct calculation. Consider two positions of the rocket as shown in Figure 8-5. Let $\Delta t^* = t^*_2 - t^*_1$. Let Δt be the time interval on earth between seeing t^*_2 and t^*_1 through the telescope. From the earth's viewpoint, the *actual* time that elapses between the two readings on the rocket clock is $\Delta t^*/\sqrt{1 - V^2/c^2}$ (rocket clock is running slow). During this time interval the rocket travels further away by an amount $V\,\Delta t^*/\sqrt{1 - V^2/c^2}$ So the t^*_2 reading is delayed from reaching the earth by an extra amount of time equal to

$$\frac{\text{Extra distance}}{c} = \frac{V}{c}\,\frac{\Delta t^*}{\sqrt{1 - V^2/c^2}}$$

Hence the total time that elapses on the earth between seeing t^*_1 and t^*_2 is

$$\Delta t = \frac{\Delta t^*}{\sqrt{1 - V^2/c^2}} + \frac{V}{c}\,\frac{\Delta t^*}{\sqrt{1 - V^2/c^2}} = \frac{\Delta t^*(1 + V/c)}{\sqrt{1 - V^2/c^2}}$$

or

$$\Delta t = \frac{\Delta t^*(1 + V/c)}{\sqrt{1 - V^2/c^2}} \qquad (8\text{-}1)$$

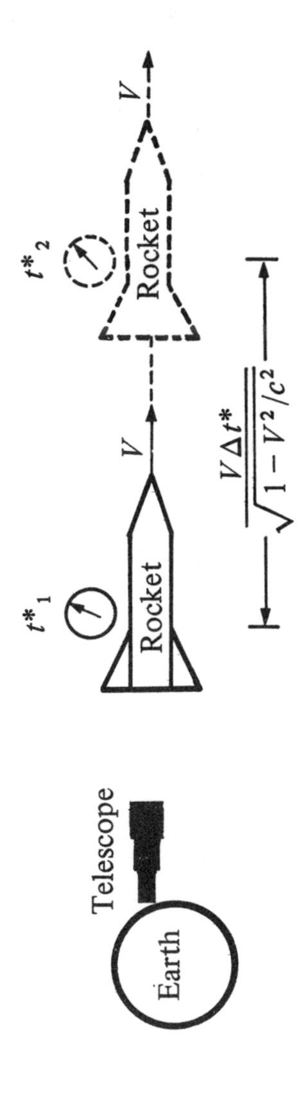

FIGURE 8–5:
Earth Observer Viewing the Rocket Through a Telescope

For future reference, we note here that if the rocket were traveling toward the earth, we would subtract the

$$\frac{V}{c}\frac{\Delta t^*}{\sqrt{1 - V^2/c^2}}$$

instead of adding it, and

$$\Delta t = \frac{\Delta t^*(1 - V/c)}{\sqrt{1 - V^2/c^2}} \tag{8-2}$$

Equations (8-1) and (8-2) express one form of what is known as the relativistic Doppler effect (see Question 14-3 in Chapter 14). In both equations we assume V to be a positive number. If we agree to let V be positive if the clock is approaching the telescope and to let V be negative if the clock is receding from the telescope, then equation (8-2) will apply to both situations.

In our problem, with $V = (4/5)c$, equation (8-1) becomes

$$\Delta t = \Delta t^* \frac{9/5}{3/5} = 3\Delta t^*$$

So $\Delta t^* = \Delta t/3$, which confirms the factor of 1/3 in the rate at which the rocket clock appears to run.

In principle, the earth observer could measure how far away the rocket clock was when it emitted the light the earth observer sees later in the telescope. This could be done by determining the angle subtended at the telescope by the rays coming from the edges of the rocket clock. This information, along with a knowledge of the width of the clock, would uniquely determine its position when it emitted the light. By this method the earth observer could confirm that the clock was 8 light-years away when it emitted the light seen through the telescope 18 years after the rocket left the earth.

After 18 years the earth observer begins to see the light emitted by the rocket clock during its return journey. All the light emitted during the return trip reaches the earth during the next 2 years. The rate at which the rocket clock appears to run during these 2 years is given by equation (8-2).

$$\Delta t = \frac{\Delta t^*(1 - V/c)}{\sqrt{1 - V^2/c^2}} = \Delta t^* \frac{1/5}{3/5} = \frac{\Delta t^*}{3}$$

or $\Delta t^* = 3\Delta t$. Thus the rocket clock appears to run fast by a factor of 3 during these 2 years. Therefore, it ticks off 6 more years during the return trip, and should read 12 years when it gets back. Through the telescope it appears to travel the 8 light-years in only 2 years, so its *apparent* approaching speed is $4c$ (as measured by the rate at which its angular size increases in the telescope). Of course, if the earth observer corrects his observations by allowing for the time lag for the light emitted by the rocket clock to reach the telescope, he computes the *true* speed of the rocket to be $4c/5$.

In summary, then, from the data obtained by the earth-bound observer looking through the telescope, he sees the rocket clock apparently running slow by a factor of 1/3 for 18 years and then running fast by a factor of 3 for 2 years. Again, the earth's clock reads 20 years and the rocket clock reads 12 years when they are back together.

Analysis of Journey from Data Obtained by a Rocket Observer Looking through a Telescope

Method 5:

Analysis of the round trip in terms of the data obtained by the rocket observer viewing the earth's clock through a telescope.

We assume that the earth's clock is continually observed through a telescope by the observer on the rocket. The rocket observer keeps a record of what he sees through the telescope at various times by his own clock. Let t^* be the time on the rocket observer's clock and let t be the time *seen* on the earth's clock by the rocket observer at time t^*.

Equations (8-1) and (8-2) apply to the case if we interchange Δt and Δt^* in these equations (because we have interchanged the roles of the unstarred and the starred times).

Earth clock receding from the rocket's telescope:

$$\Delta t^* = \frac{\Delta t\left(1 + \frac{V}{c}\right)}{\sqrt{1 - V^2/c^2}} \tag{8-4}$$

Earth clock approaching the rocket's telescope:

$$\Delta t^* = \frac{\Delta t\left(1 - \frac{V}{c}\right)}{\sqrt{1 - V^2/c^2}} \tag{8-5}$$

When the rocket observer's clock reads 6 years, he is arriving at the star. During these 6 years the earth's clock has appeared through the telescope to be running slow by a factor of 1/3, so at the instant the rocket is arriving at the star the rocket observer sees the earth clock reading 2 years and it appears to be 8/3 light-years away. Now the rocket suddenly turns around and begins to approach the earth. Just after the turnaround, the observer still sees the earth's clock reading 2 years but its apparent distance away is 24 light-years! Of course, the earth didn't suddenly change its position, but when the rocket changed from one inertial frame to another, the apparent angle subtended by the light rays from the earth's clock changed by a factor of 1/9, so the earth's clock appears nine times farther away. (This result is derived at the end of the chapter.) As in Method 4, now that the rocket is observing light from an approaching earth clock, the earth clock appears to be running fast by a factor of 3 and approaching the rocket at an apparent speed of $4c$. Therefore, 6 more years should elapse before the earth clock reaches the rocket (it must travel 24 light-years at a speed of $4c$). This implies that the rocket clock should read 12 years when the earth and the rocket are back together. The earth clock appeared to run fast by a factor of 3 during the last 6 years, so it ticked off 18 years (through the telescope) during this phase. But it read 2 years through the telescope at the turnaround instant. Thus the earth's clock should read a total of 20 years at the end. In summary, the earth's clock ran slow by a factor of 1/3 for 6 years and fast by a factor of 3 for 6 years, as seen through the telescope by the rocket observer. Again the final clock readings are in agreement with those predicted by the other methods.

Finally, let us confirm our statement that the apparent distance of the earth's clock from the rocket as seen through the telescope changes from 8/3 light-years to 24 light-years during the turnaround. From the rocket viewpoint before turnaround (i.e., from the viewpoint of the starred inertial frame used in Method 2), the earth's clock would read 2 years at an actual time of $2 \times 5/3 = 3^1/_3$ years after the rocket left the earth. But the separation after that time would be ($v/c \times 3^1/_3$) light-years = $4/5 \times 3^1/_3$ = 8/3 light-years. The angle subtended by the clock's rays should yield this value. However, after the turnaround, the distances computed by the subtended angle should be consistent with the view of the primed inertial frame (the one approaching the earth at a speed of $4c/5$). The light observed by the rocket at the star just after turnaround left the earth's clock when it read 2 years. We then ask the primed frame the following question: How far had the light emitted by the earth's clock when it read 2 years traveled before it reached the star?

We can refer to the analysis in Method 3 to help us arrive at our answer. Our discussion is in terms of the viewpoint of the primed frame. The rocket traveled 82/3 years to catch the star. Its speed was $(40/41)c$. Thus the distance from the initial position of the earth to the position of the star when the rocket caught up to the star must be $40/41 \times 82/3 = 80/3 = 26\,{}^2\!/_3$ light-years.

When the earth's clock reads 2 years, a time of $5/3 \times 2 = 10/3$ years has elapsed. The distance of the earth from its initial position is $V/c \times 10/3 = 4/5 \times 10/3 = 40/15 = 2\,{}^2\!/_3$ light-years. Therefore, its distance from where the rocket catches up to the star is $26\,{}^2\!/_3 - 2\,{}^2\!/_3 = 24$ light-years. The angle subtended by the light rays reaching the rocket immediately after the turnaround must be consistent with this distance.

Suggested practice problems to ponder in Appendix A: P-3(c), (d); P-7; P-11; P-12; P-13(a); P-15(d); P-18(a), (b).

CHAPTER 9

The Earth as a Rocket Ship

In this chapter we consider a variation of the situation analyzed in Chapter 8.

The Earth Accelerates!

Suppose a rocket ship starts out from the earth moving at a constant velocity of $4c/5$ relative to the earth. After the rocket ship starts on its journey, workers back on earth construct a giant rocket motor. They fasten the rocket motor securely to the earth. Exactly 10 years after the first rocket ship left the earth, the giant rocket motor is fired and the entire earth is propelled with great acceleration toward the first rocket ship. In a short time the earth reaches a velocity high enough so that from the viewpoint of the first rocket ship the earth is gaining on the rocket ship with a relative speed of $4c/5$. Assume that the acceleration time of the earth in reaching its final velocity is a negligible fraction of a year.

Question 9-1:

What will be the readings on the earth clock and on the rocket ship clock when the earth catches up with the rocket ship?

This problem could be analyzed correctly from the viewpoint of any inertial frame. The earth changes its inertial frame during the process, so one would have to be careful in trying to deduce the answer from the earth observers' viewpoint. The situation is easiest to analyze from the viewpoint of the inertial frame of the rocket ship. Let us describe the entire process from this viewpoint.

Rocket Ship Observers' Explanation

At first the earth is traveling at a speed of $4c/5$ away from the rocket ship. When the earth's clock reads 10 years, the earth suddenly accelerates and begins to come back toward the rocket ship with a relative speed of $4c/5$. The earth's clock runs slow by a factor of 3/5 during the journey. If it ticked off 10 years during the outward portion of the trip, the actual time for this portion was $10 \times 5/3 = 16^2/_3$ years. The earth then came back at the same relative speed, so it would take another $16^2/_3$ years to come back, and the earth's clock would tick off another 10 years. When the earth got back, its clock would read 20 years and the rocket ship clock would read $33^1/_3$ years.

Question 9-2:

What was the maximum separation of the earth and the rocket ship, according to the rocket ship observers?

The earth traveled outward at a speed of $4c/5$ for $16^2/_3$ years, according to the rocket ship observers. So the maximum distance between the two must have been $(4/5) \times 16^2/_3 = 13^1/_3$ light-years.

Question 9-3:

What was the separation between the rocket ship and the earth just before the earth accelerated, according to the earth observers?

The rocket ship had been traveling for 10 years at a speed of $4c/5$, according to the earth's observers. So its distance away was $(4/5) \times 10 = 8$ light-years.

Question 9-4:

What was the *final* speed of the earth relative to its *initial* inertial frame?

The speed of the rocket ship relative to the initial inertial frame of the earth was $4c/5$. The *final* speed of the earth relative to the rocket ship was $v^* = 4c/5$. So putting $V = 4c/5$ and $v^* = 4c/5$ into the addition law of velocities [equation (6-3)] yields the value of the final speed of the earth relative to the earth's initial inertial frame.

$$v = \frac{v^* + V}{1 + \dfrac{v^* V}{c^2}} = \frac{4c/5 + 4c/5}{1 + (4/5) \times (4/5)} = \frac{8c/5}{41/25} = 40c/41$$

Suggested practice problems to ponder in Appendix A: P-6 (a), (b), (c), (d).

CHAPTER 10
Squeezing a Long Pole Into a Short Barn

A very smart farmer owned a barn. It was 35 feet long, and it had a small door at each end. Next to the barn was a 40-foot iron pole. Whenever the farmer put the pole lengthwise in the barn, 5 feet of the pole's length stuck out through the doors and it was not possible to shut both doors at the same time. Figure 10-1 shows a top view of the situation.

A traveling salesman stopped by to sell the farmer some merchandise. He noticed the pole in the barn, and remarked how unfortunate it was that both barn doors couldn't be closed with the pole inside. The farmer answered that he knew of a way to get the barn doors closed with the pole inside, and he wouldn't have to cut, bend, or tip the pole or alter the barn in order to do it. The salesman offered to give the farmer some of his wares in return for being let in on the farmer's secret.

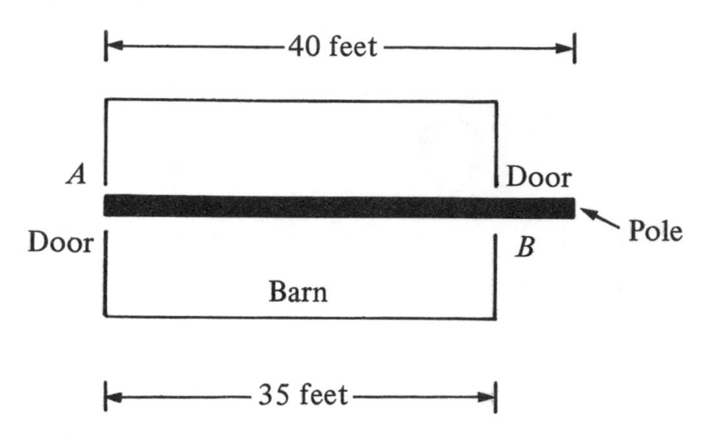

FIGURE 10–1:
The Pole and the Barn

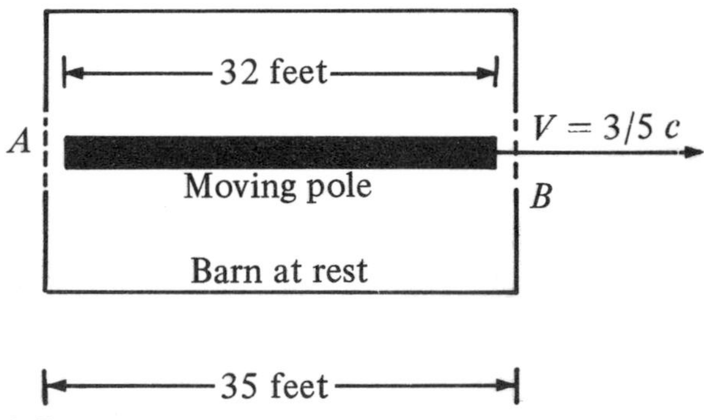

FIGURE 10–2:
Moving Pole Inside the Barn (Barn Frame's Viewpoint)

Question 10-1:

What procedure did the farmer have in mind?

The farmer had studied relativity theory. He knew that if the pole could be accelerated to a speed of $3c/5$ relative to the barn, it would automatically undergo the Lorentz contraction and its length would be $(4/5) \times 40 = 32$ feet.

Then it would fit into the barn. For a short time interval both doors could be closed at the same time with the pole inside. To keep the pole from bumping into one of the doors, one of the barn doors must be reopened immediately; nevertheless, for a while at least, the pole would be inside the barn with the doors closed. Figure 10-2 shows the moving pole inside. The diagram is drawn from the viewpoint of the barn's inertial frame.

Question 10-2:

> Let us look at the situation from the viewpoint of an observer in an inertial frame moving to the right with a speed of $3c/5$ relative to the barn. In the final situation the pole is at rest in this frame and the barn is moving. The pole length would be 40 feet and the barn length would be $(4/5) \times 35 = 28$ feet. How does this observer describe the situation presented in Question 10-1?

Because of the disagreement on simultaneity, according to the pole inertial frame the two doors were not closed at the same time. Door B in Figure 10-2 was closed first (see Chapters 2 and 5), and then was reopened. Figure 10-3 shows the instant door B was closed, according to the pole observer. Later, door B is reopened and later yet, door A is closed as shown in Figure 10-4.

The observer at rest with respect to the pole says that the pole never was completely inside the barn. Door B was closed while the left end of the pole was still outside the barn, and then door A was closed after the left end had passed through it.

Suggested practice problem to ponder in Appendix A: P-19.

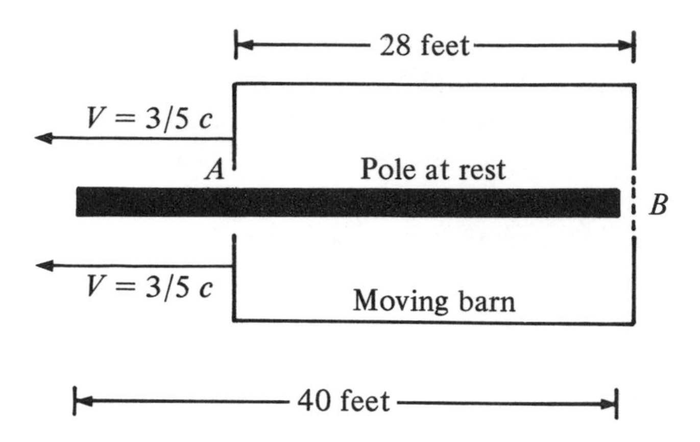

FIGURE 10-3:
Situation at the Instant Door B was Closed (Pole Frame's Viewpoint)

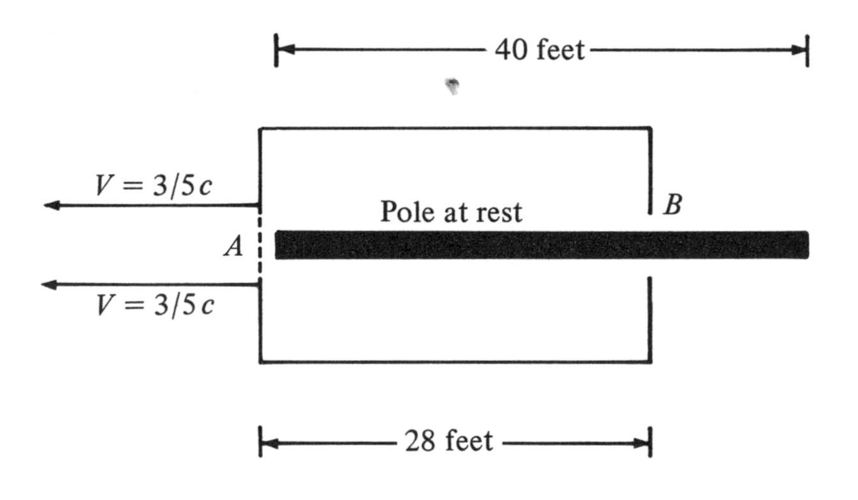

FIGURE 10-4:
Situation at the Instant Door A was Closed (Pole Frame's Viewpoint)

CHAPTER 11

"Follow-the-Leader" with Two Fast Rockets

Suppose we are observers at rest in some inertial reference frame. George and his observers are in another inertial reference frame, which is moving toward the right at a speed of $4c/5$ relative to us.

Two identical rockets are initially at rest in our frame. Their dimensions and positions are shown in Figure 11-1 (a). All lengths are specified in meters.

In a shirt pocket of an astronaut on one of the rockets is a string that is capable of being stretched to a length of 300 meters, but it will break if stretched beyond 300 meters.

Now suppose the two identical rockets are fired simultaneously (according to us) so that each one travels toward the right in a straight line. After the firing is completed the rockets are both traveling toward the right at a speed of $4c/5$ relative to us. Therefore, the rockets will end up *at rest* in George's reference frame.

Question 11-1:

Draw a diagram showing the *final* situation from our viewpoint. Explain your diagram.

Figure 11-1(b) gives the diagram asked for. The nose-to-nose separation remains 200 meters throughout the acceleration process because the rockets are identical and were fired simultaneously. In fact, the distance between *every* pair of corresponding points on the two rockets remains constant at 200 meters. However, the rockets themselves undergo the Lorentz contraction, and so end up with a length of 60 meters.

Question 11-2:

Draw a diagram showing the initial situation from the viewpoint of George's reference frame. Explain your diagram.

Figure 11-1(c) is the correct diagram. It is obtained from Figure 11-1(a) by applying the Lorentz contraction factor of 3/5 to all the designated distances.

Question 11-3:

Draw a diagram showing the *final* situation from the viewpoint of Goerge's reference frame. Explain your diagram.

Figure 11-1(d) is the diagram for the final situation from George's viewpoint. In the final situation the rockets are at rest in George's frame. Consequently, Figure 11-1(d) was obtained by computing the *rest* lengths and separations that are compatible with the *moving* lengths shown in Figure 11-1(b).

Note that Figure 11-1(c) is merely the Lorentz-contracted form of Figure 11-1(a), whereas Figure 11-1(b) is the Lorentz-contracted form of Figure 11-1(d).

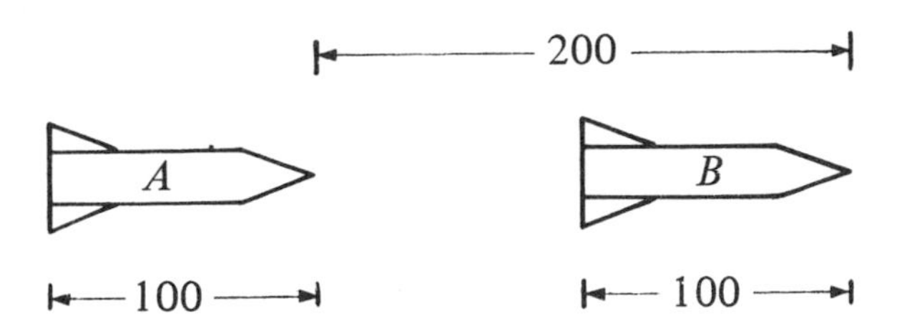

FIGURE 11-1(a):
Initial Situation from Our Viewpoint

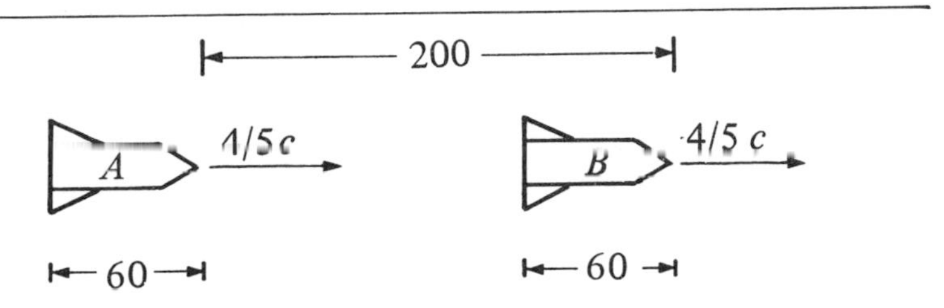

FIGURE 11-1(b):
Final Situation from Our Viewpoint

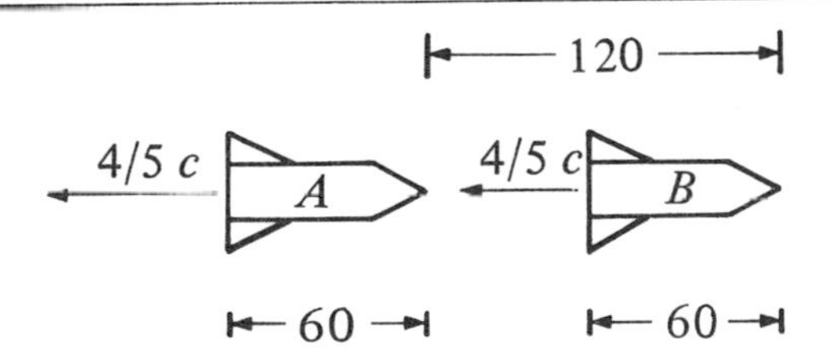

FIGURE 11-1(c):
Initial Situation from George's Viewpoint

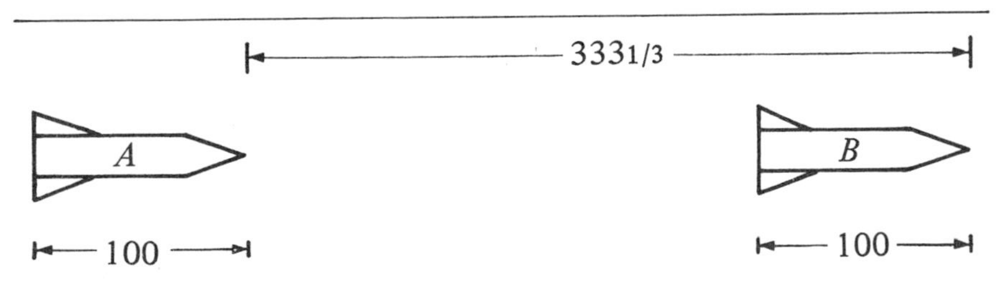

FIGURE 11-1(d):
Final Situation from George's Viewpoint

George's Explanation

Question 11-4:

How does George explain the transition between the initial and final situations?

According to George, initially both rockets were traveling to the left at a speed of $4c/5$. The initial nose-to-nose separation was 120 meters. Each rocket comes to rest after it is fired. But rocket B is fired first, according to George. (Remember that the firing, although simultaneous to us, was *not* simultaneous to George's observers.) Rocket B stops first, and rocket A continues on to the left for a short time. When A does stop, the nose-to-nose separation has increased to 333.33 meters from the original value of 120 meters.

It is instructive to make an independent calculation of the final separation. According to the results in Chapter 5, George believes our clocks are out of synchronization, such that a clock on our left rocket in Figure 11-1(c) would be LV/c^2 behind a clock on the right rocket. So he says our left clock has to tick off that amount after rocket B has stopped and before rocket A stops. But he also says the clock runs slow, so according to him the actual time lapse involved is $(VL/c^2)/\sqrt{1-V^2/c^2}$. During this time the left rocket coasts a distance equal to

$$V \times (\text{time lapse}) = \frac{V^2 L}{c^2} / \sqrt{1-V^2/c^2}$$

Putting in numbers gives $(4/5) \times (4 \times 5) \times (5/3) \times 200 = 213.33$ meters. Add this to the 120-meter separation, according to George, that existed before the rockets stopped. This gives the total separation at the end as 333.33 meters.

Question 11-5:

If during the final situation the astronaut climbs out of his rocket and tries to attach his string between the noses of the two rockets, will he be able to do so without breaking the string?

No. The string would break. It would have to stretch to a length of 333.33 meters in order to connect the two noses.

Suggested practice problems to ponder in Appendix A: P-14; P-17(f).

CHAPTER 12

Slipping Through the Net

A wire net was constructed with fine wires spaced 10 cm (centimeters) apart. A steel rod with a rest length of 15 cm passes over the net with a speed of $4c/5$ relative to the net. Its moving length is $(3/5) \times 15 = 9$ cm because of the Lorentz contraction. See Figure 12-1.

Suppose a series of fingers grabs the rod simultaneously at various positions along its length and yanks it down through the net, while letting it maintain its component speed of $4c/5$ toward the right. The rod should be able to pass through the mesh without striking one of the wires. Figure 12-2 shows the sequence of positions of the rod as it passes through, from the viewpoint of the net inertial reference frame. All this is perfectly reasonable from the viewpoint of the net inertial reference frame.

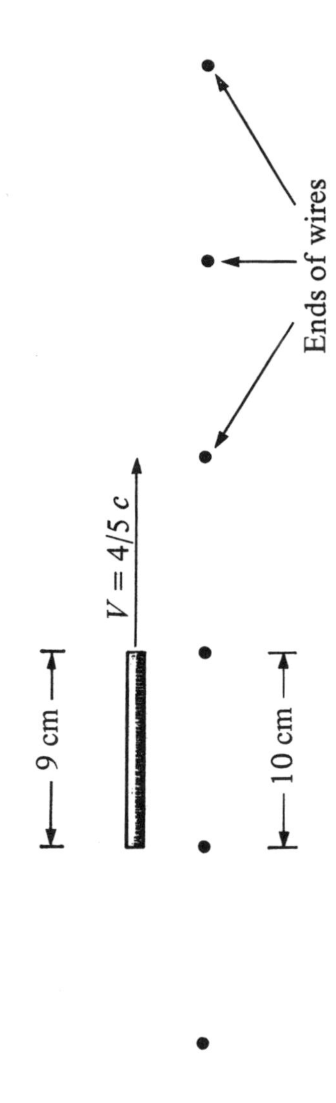

FIGURE 12–1:
Edge View of Wire Net with Rod Passing Over It (Net Frame's Viewpoint)

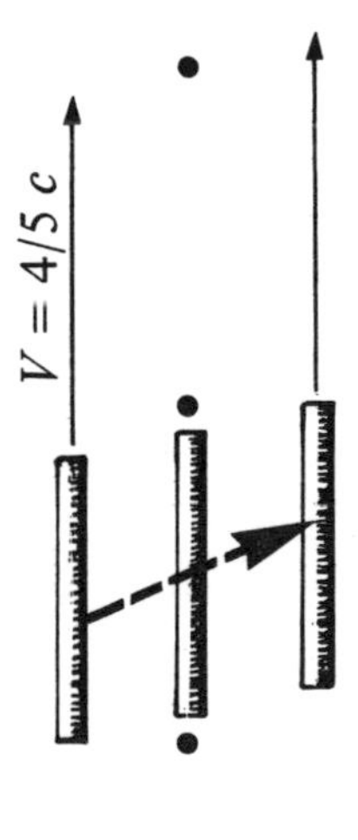

FIGURE 12-2:
Rod Passing Through the Net (Net Frame's Viewpoint)

Question 12-1:

What is the description of the preceding process from the viewpoint of an inertial frame of reference moving to the right at a speed of $4c/5$ relative to the net?

The rod is at rest in this reference frame, and the net is moving to the left. The rod is 15 cm long and the moving spacing between the wires is $(3/5) \times 10 = 6$ cm. The initial situation appears to the rod reference frame as shown in Figure 12-3.

How did the rod get below the net? All parts of the rod were yanked through simultaneously from the viewpoint of the net reference frame. Again, because of the disagreement on simultaneity, the right end was yanked first according to the rod reference frame. Then later the other portions were yanked through. From the rod reference frame's view, the rod slithered through the mesh, as illustrated by the time sequence of situations in Figure 12-4.

This problem calls attention to another prejudice that must be forsaken. According to old ideas, if a rod moved so that it was always straight from the viewpoint of one reference frame, then it would also always be straight from the viewpoint of any other reference frame. In other words, whether or not a rod was straight was supposed to be an absolute property of the rod and not of the reference frame. However, the new concepts in relativity theory change this. An accelerated rod (in this problem it was accelerating when it was yanked through the net) may be a straight rod at all times in one frame but appear to be crooked or bent from the viewpoint of another reference frame. This is merely another consequence of the disagreement on simultaneity.

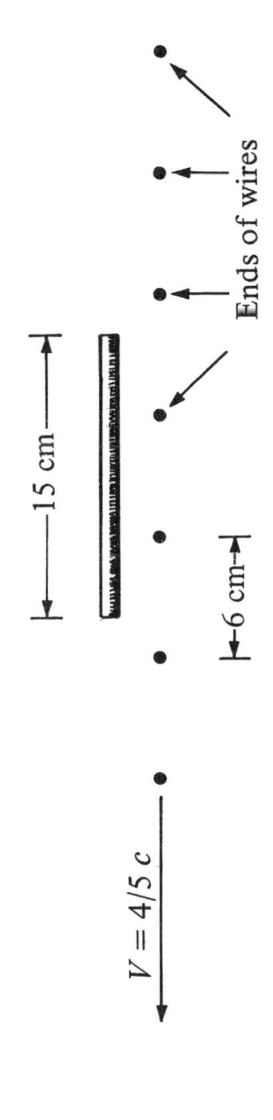

FIGURE 12-3:
Net Passing Under the Rod (Rod Frame's Viewpoint)

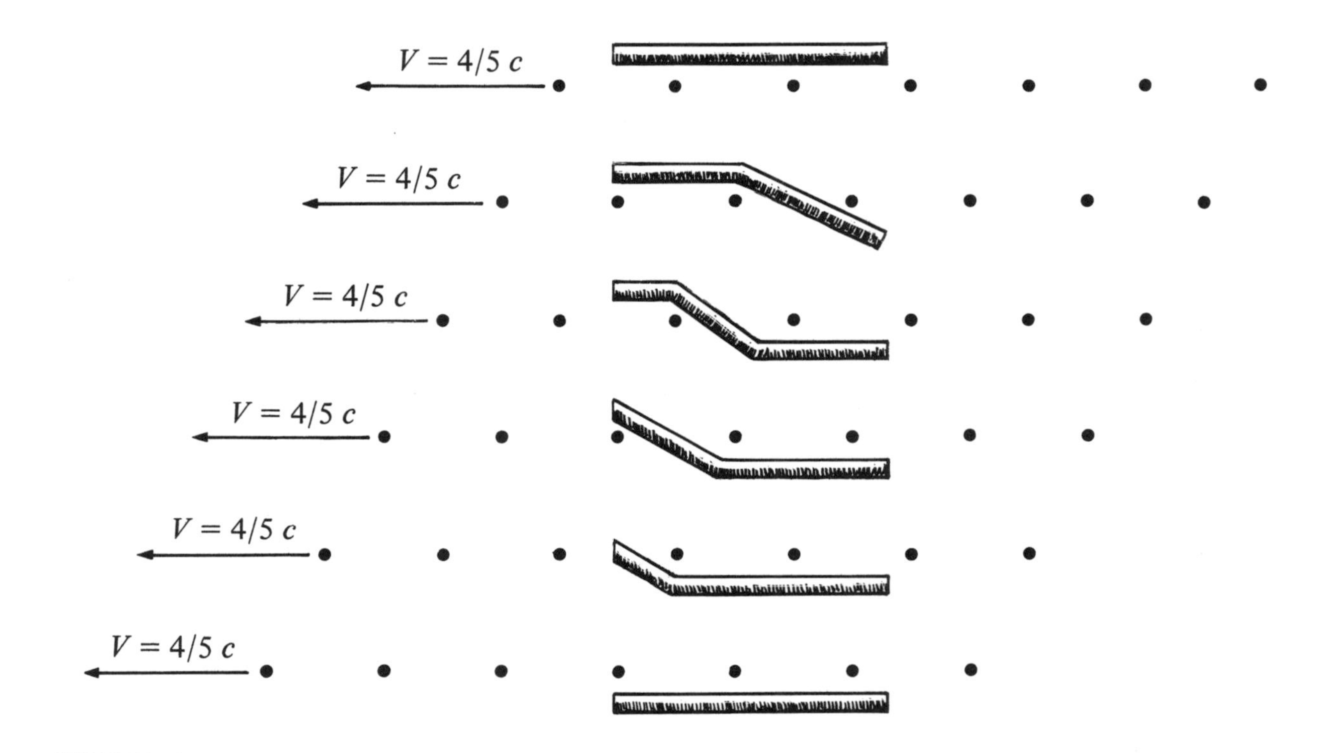

FIGURE 12–4:
Series of Pictures of Rod Passing Through the Net (Rod Frame's Viewpoint)

Beware the "Rigid" Object in Relativity Theory

The concept of the rigid rod, a very natural and useful idealization in prerelativity physics, has almost no validity in relativity. In fact, no generally satisfactory and useful definition for rigidity, which is compatible with relativity, has yet been found. The usual nonrelativistic definition of a rigid rod implies that if you push on one end, the other end immediately moves also. Such behavior would enable a person to use the rod to send signals infinitely fast, and most certainly is incompatible with relativity. The author has proposed a definition that is appropriate to the very special case of the straight-line translational motion of a thin rod.[1] The definition was derived on the basis that the rod cannot be used to send signals faster than the speed of light and that the definition must have the same form in all inertial reference frames.

Many paradoxes can arise if one inadvertently carries the old idea of rigidity over into a relativistic situation. The reader is warned to remain alert to avoid this difficulty.

---

1. Evett, Arthur A., *Nuovo Cimento*, vol. 68, 1963, pp. 685–694; 706–711.

CHAPTER 13

Snapshots of a Moving Reference Frame

In this chapter we are concerned with two inertial reference frames. First we set up a diagram for "our" inertial reference frame, which we refer to as the unstarred reference frame. To do this we mark off our x-axis in feet and place a set of observers with synchronized clocks at various fixed locations on the x-axis. The hand on each of these special clocks takes 2 μs to complete one sweep around the clock's face. The lower portion of Figure 13-1 shows the situation in our frame from our view at the instant $t = 0$. The space interval between adjacent displayed clocks is 800 feet. The speed of light is considered to be exactly $c = 1000$ ft/μs.

Now consider the *starred* inertial reference frame, which is moving at a speed of $V = 4c/5$ relative to us in the positive x direction. Observers at rest in this frame have marked off their x^*-axis in feet and they have their set of synchronized clocks placed at various fixed locations on the x^*-axis. They, *in their frame,* have done exactly what we did in our frame.

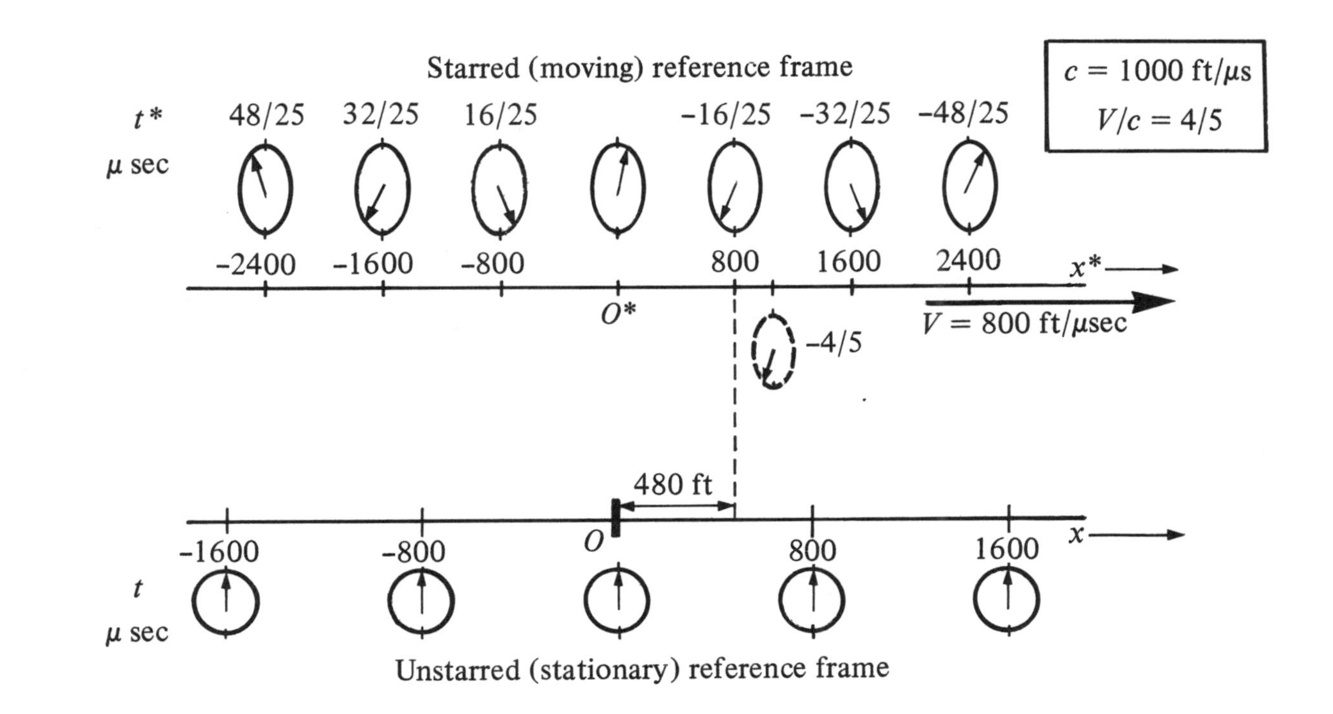

FIGURE 13-1:
Situation at the Instant $t = 0$ from the Unstarred View

Any one of the observers fixed in the unstarred frame can analyze what goes on in his frame by gathering together information from all the other observers in his frame. Each observer reports on events occurring at his own location only (by specifying his x location and the reading on his clock for each significant event at his location). This information can all be reported to one of the observers, who then can plot it in a composite diagram showing the total situation from the viewpoint of his reference frame at any instant. A similar procedure could be used by the set of observers at rest in the starred frame to put together their composite of what is happening.

Agreement on Raw Data, Disagreement on Interpretation

A possible event is one of the starred observers passing one of the unstarred observers. The starred observer will record his x^* and t^* values for the event, and the unstarred observer will record his x and t values for the same event. Also, each one can look at the other as they pass and can observe what the other is recording. These x, t, x^* and t^* values are the raw data for the events. The starred values can be computed from the unstarred values by using the Lorentz transformation equations. There is no dispute between the two reference frames as to the validity of the raw data (because each one can observe the other), but they may have widely differing ideas on the *interpretation* of the raw data for some sequence of events. Each should be able to explain not only the raw data taken from his own set of observers for the sequence of events, but also the raw data of the observers on another reference frame for the same sequence of events.

Figure 13-1 shows a plot of events on the x- and x^*-axes, with all the events at $t = 0$. This corresponds to one single instant of time from the viewpoint of the unstarred reference frame, so the composite diagram represents simultaneous events from this viewpoint. As the t^* values for these events are not identical, the composite represents a plot of events at different times from the view of the starred observers. Consequently, the diagram in Figure 13-1 has a simple direct significance only to the unstarred frame. A diagram could equally well be drawn that would display all events corresponding to some given value of t^*; such a diagram would represent events simultaneous in the starred frame.

For simplicity, it is assumed that the event $x = 0$, $t = 0$ corresponds to $x^* = 0$, $t^* = 0$. This means it has been arranged for the origin clocks in the two frames both to read zero when they pass each other.

Interpreting Figure 13-1 as a Photograph

Figure 13-1 could be the print of a photograph taken in a particular way. Suppose a *long* photographic plate were placed along our x-axis. Each small portion of the long plate is covered by a shutter, and when the shutter at one point is opened, only light directly out from the plate at that location falls on the plate. Our observers at different locations along the plate have instructions to open their shutter for a very short time at time $t = 0$. Thus the entire plate gets exposed simultaneously, according to us. The plate would show the starred clocks as well as ours at $t = 0$. Figure 13-1 would be the picture appearing on the plate when it was developed.

Similarly, the starred frame could get a picture corresponding to the situation at $t^* = 0$. A long plate would be placed at rest along that frame and the observers would expose all parts of the plate at $t^* = 0$. Their picture would be different from ours, because, according to us, they exposed different portions of the plates at different times. Figure 13-2 shows their picture.

How the Numbers Were Obtained for Figure 13-1

Let us examine how Figure 13-1 was constructed. The portion for the unstarred frame is self-evident. The portion showing the starred frame will appear to be Lorentz-contracted (from the unstarred view), so the starred clocks will be $800 \times \sqrt{(1 - V^2/c^2)} = 800 \times 3/5 = 480$ feet apart. This spacing is indicated on the diagram. Each of the starred clocks is shown to be elliptical rather than circular, emphasizing that it too suffers a Lorentz contraction parallel to the direction of relative motion. (It should be remembered that the diagram shows the situation at *one instant* from the viewpoint of the unstarred frame. Each observer in the unstarred frame records the information in his neighborhood, making allowances for any possible time lags between when he sees an event and when the event actually occurs. It is only on this basis that the moving clocks exhibit the usual Lorentz-contraction effect. If one records the appearance of a moving clock without making the light travel-time allowance, the apparent shape is not as shown in the diagram. Terell[1] and Weisskopf[2] discuss this point further.) The readings on the starred clocks are obtained from one of the Lorentz transformation equations (refer to equation 7-7):

$$t^* = (t - Vx/c^2) \,/\, \sqrt{1 - V^2/c^2} \qquad (13\text{-}1)$$

$$x^* = (x - Vt) \,/\, \sqrt{1 - V^2/c^2} \qquad (13\text{-}2)$$

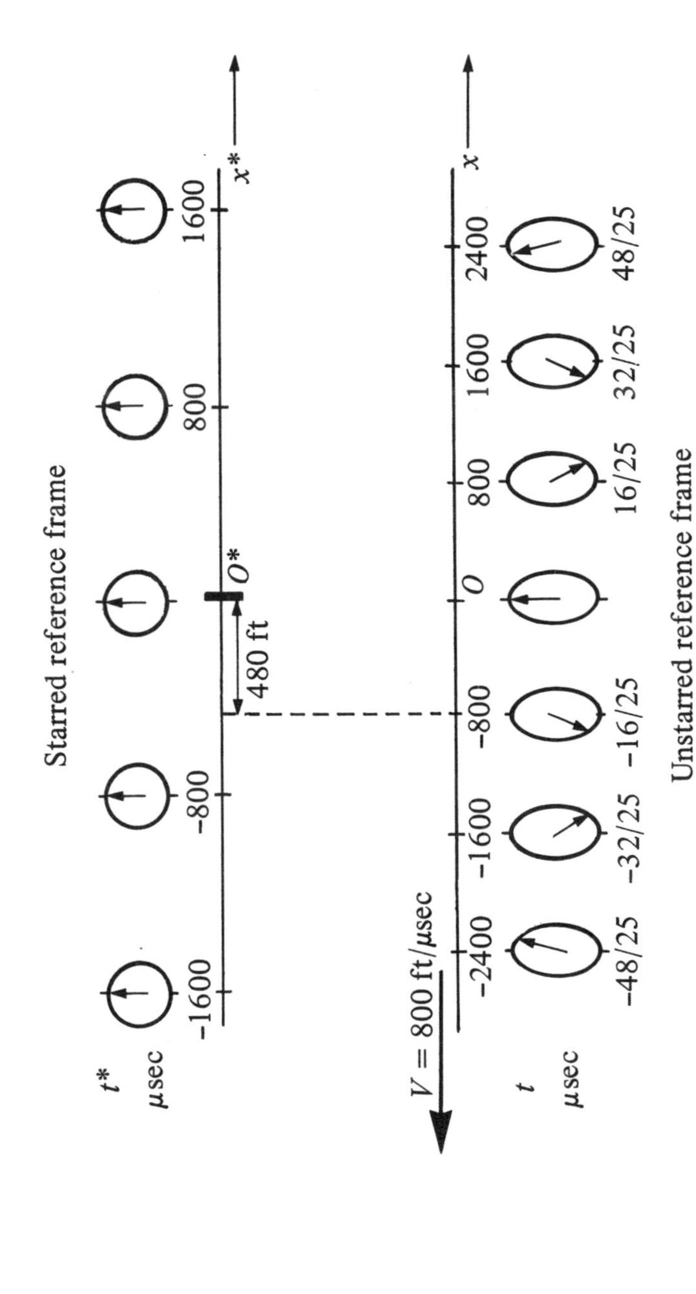

FIGURE 13–2:
Situation at the Instant $t^* = 0$ *from Starred View*

Calculation of Starred-Clock Readings

To get the t^* reading of the $x^* = 800$ clock, in equation (13-1) we substitute $t = 0$, $V/c = 4/5$, $x = 480$, $\sqrt{1 - V^2/c^2} = 3/5$. This gives $t^* = -16/25\ \mu s$. Similar calculations for the other starred clocks confirm that each clock reads 16/25 μs less than the clock behind it.

Because of the time dilation phenomenon, each starred clock runs slow by a factor of 3/5, according to the unstarred observers. If we wish to construct a diagram showing the situation at some later time $t = T\ \mu s$ (from the unstarred viewpoint), we should show the unstarred clocks all reading $t = T$, we translate the entire starred frame to the right a distance of $VT = (800 \times T)$ feet, and we should add $3T/5\ \mu s$ to the reading of each of the starred clocks shown in Figure 13-1.

Question 13-1:

> We (on the unstarred frame) watch the starred frame measure the relative speed of our frames. How do we explain their data?

The speed of the starred frame relative to the unstarred frame was specified to be 800/ft/μs. This means that the $x^* = 0$ clock would be opposite the $x = 800$ clock when the $x = 800$ clock reads 1 μs. In fact, these data would be used by the unstarred frame to deduce that the relative speed is actually 800 ft/μs.

Suppose the starred frame was asked to measure the speed of the unstarred frame relative to the starred frame. Presumably the starred frame would focus attention on one of the clocks fixed in the unstarred frame and would measure its speed. Assume that the $x = 0$ clock is the one observed. It is at $x^* = 0$ at $t^* = 0$. What will the t^* reading be when the $x = 0$ clock is opposite the $x^* = -800$ clock?

We can use the Figure 13-1 diagram to determine this information. According to the unstarred frame, the $x^* = -800$ clock reads $t^* = 16/25$ μs and is 480 feet to the left of $x = 0$ at $t = 0$. It will take (480/800) μs for this clock to get to $x = 0$. But the starred clock will tick off only 3/5 of this amount, or $(3/5) \times (3/5) = (9/25)$ μs, during this time. Hence, when the $x^* = -800$ clock gets opposite the $x = 0$ clock, the $x = 0$ clock will read 3/5 μs and the $x^* = -800$ clock will read $(16/25) + (9/25) = 1\,\mu$s. Note that we used the viewpoint of the unstarred frame in reaching this conclusion. But, as we emphasized earlier, neither frame disputes the raw data, so the starred frame must agree with the unstarred frame on the readings of the $x = 0$ and the $x^* = -800$ clocks as they pass each other. Figure 13-3 shows the overall situation at $t = 3/5$ μs from the unstarred view.

How do the starred observers interpret these raw data? They say the $x = 0$ clock was at $x^* = 0$ at $t^* = 0$ and was at $x^* = -800$ feet at $t^* = 1\,\mu$s. They conclude that the $x = 0$ clock traveled 800 feet to the left in 1 μs, which implies that the unstarred frame is traveling to the left at a speed of 800 ft/μs.

Thus both frames agree on the relative speed. Also, both frames agree on the raw data needed for each to come to this conclusion. However, there is disagreement on the correct interpretation of the processes involved in yielding the raw data. The unstarred frame argues that the $x^* = -800$ clock reads $1\,\mu$s when it is opposite the $x = 0$ clock only because the starred clocks are unsynchronized, are running slow, and are only 480 feet apart. The starred frame says the $x^* = -800$ clock reads $1\,\mu$s when the $x = 0$ clock is opposite it because at $t^* = 0$ the $x = 0$ clock was at $x^* = 0$, and $1\,\mu$s later it had traveled 800 feet to the $x^* = -800$ clock.

The important point in all this is that each frame can consistently interpret both its own readings for the events as well as the other frame's readings for the events.

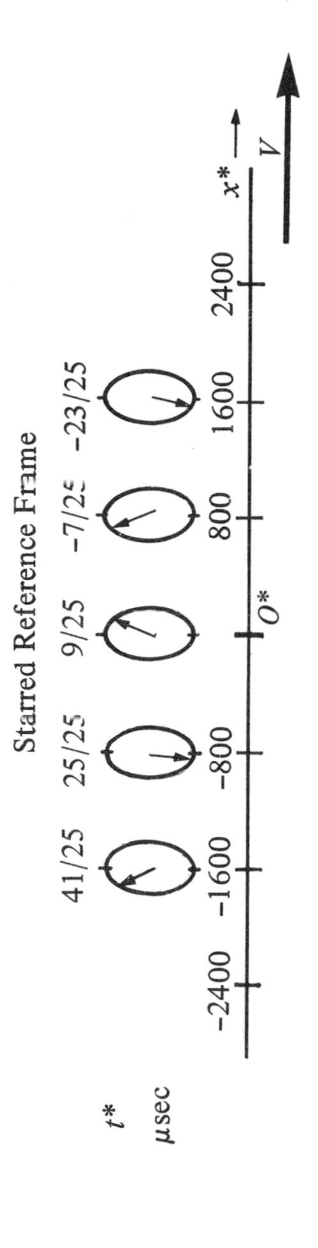

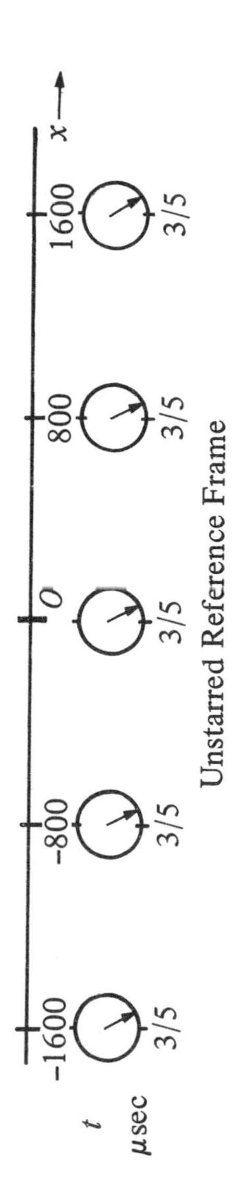

FIGURE 13-3:
Situation at the Instant t = 3/5 μsec from the Unstarred View

Question 13-2:

> We (on the unstarred frame) watch the starred frame measure the rate our clocks run. How do we explain their data?

In discussing the diagram in Figure 13-1, we have used the fact that, from the unstarred view, each of the starred clocks runs slow by a factor of 3/5. This implies that the $x^* = 0$ clock should read $3/5\ \mu s$ when it is opposite the $x = 800$ clock at $t = 1\ \mu s$. The raw data would be the basis for the unstarred frame to argue that the starred clock was running slow.

Suppose the observers in the starred frame were asked to measure the ticking rate of our unstarred clocks. They would observe one of the clocks fixed in the unstarred frame, say the $x = 0$ clock. The $x = 0$ clock reads $t = 0$ when it is opposite the $x^* = 0$ clock reading $t^* = 0$. According to the result obtained in the preceding section (see Figure 13-3) on comparing relative speeds, we know the $x = 0$ clock reads $t = 3/5\ \mu s$ when it is opposite the $x^* = -800$ clock reading $t^* = 1\ \mu s$.

Reciprocity of Time Dilation

The starred frame concludes that the $x = 0$ clock reads $t = 0$ at $t^* = 0$, but reads only $3/5\ \mu s$ at $t^* = 1\ \mu s$. So from the viewpoint of the starred frame, the unstarred clock runs slow by a factor of 3/5.

Each frame, then, is able consistently to claim that the other frame's clocks are running slow.

Question 13-3:

> We (on the unstarred frame) watch the starred frame measure the speed of light. How do we explain their data?

Suppose the unstarred frame had a synchronized clock at $x = 1000$ feet. Then, if a beam of light were started from $x = 0$ at $t = 0$, it should reach the $x = 1000$ clock when it reads $t = 1\ \mu s$. The conclusion would be that the speed of light is 1000 ft/μs.

If the starred frame watches this same light beam, it can use its own raw data to get the speed of light from its viewpoint. A beam of light could be sent toward the right from $x^* = 0$ at $t^* = 0$. The starred frame would be particularly interested in the reading on the $x^* = 1000$ clock when the beam gets to this clock. We can compute this reading by using Figure 13-1 and the unstarred frame's viewpoint. The diagram shows the $x^* = 1000$ clock at $t = 0$. It is located at $x = (1000) \times (3/5) = 600$ feet and reads $t^* = -4/5\ \mu s$.

First we compute the time t at which the light beam will catch up with the $x^* = 1000$ clock. At the end of time t the beam will be at a distance of $(1000t)$ feet from $x = 0$, and the $x^* = 1000$ clock will be $(600 + 800t)$ feet from $x = 0$. If t is the time the beam and the $x^* = 1000$ clock are at the same distance from $x = 0$, then t must satisfy $1000t = 600 + 800t$. This gives $t = 3\ \mu s$. According to the unstarred frame, then, a time interval of 3 μs elapses before the beam overtakes the $x^* = 1000$ clock. This means that each starred clock will tick off $(3/5) \times 3 = (9/5)\ \mu s$. But the $x^* = 1000$ clock read $-(4/5)\ \mu s$ at the start of the process, so it will read $t^* = -(4/5) + (9/5) = 1\ \mu s$ when the beam gets to it.

Agreement on the Speed of Light

Using the unstarred view, we have determined that the beam leaves $x^* = 0$ at $t^* = 0$ and gets to $x^* = 1000$ feet at $t^* = 1\ \mu s$. These are the raw data that are interpreted by the starred frame to mean, very simply, that the light beam travels 1000 feet in 1 μs, or that the speed of light relative to the starred frame from the starred frame's viewpoint is 1000 ft/μs.

The relative speed between the light beam and the starred frame, from the unstarred viewpoint, is 200 ft/μs. The unstarred frame argues that it takes 3 μs for the beam to gain the 600 feet required to catch the $x^* = 1000$ clock, and the $x^* = 1000$ clock reads 1 μs only because it was out of synchronization and was running slow.

Again, both frames agree on the raw data but differ on the interpretation. Each concludes, however, that the speed of light is 1000 ft/μs relative to his own frame.

Reciprocity of the Lorentz Contraction

Question 13-4:

> We (on the unstarred frame) watch the starred frame measure the distance between our adjacent clocks. How do we explain their conclusion that our clocks are closer together than theirs?

Figure 13-1 is drawn showing that the starred frame is Lorentz-contracted by the factor of 3/5 from the unstarred view. As a consequence of this contraction, if the $x = 0$ and the $x = 800$ observers both looked up simultaneously at $t = 0$, the $x = 0$ observer would see the $x^* = 0$ clock opposite him, and the $x^* = 800$ clock would be to the left of the $x = 800$ observer. It is this type of observation that would lead the unstarred observers to conclude that the spacing of the starred clocks is less than 800 feet.

Now suppose the starred observers are interested in determining qualitatively the spacing of the unstarred clocks. The starred observers could all look out at $t^* = 0$ (which would be simultaneous from their view) and then interpret their observations.

The $x^* = 0$ observer would look out at $t^* = 0$ and would see the $x = 0$ clock opposite him. But according to Figure 13-1, the $x^* = 800$ clock reads $t^* = -16/25\ \mu s$, so he wouldn't be looking out yet (from the unstarred view). A time t would elapse equal to $(5/3) \times (16/25) = 16/15\ \mu s$ (remember the starred clock is running slow according to the unstarred view) before the $x^* = 800$ clock would read $t^* = 0$. During this time the $x^* = 800$ observer would pass to the right side of the $x = 800$ clock. So when the $x^* =$ 800 observer does look out when his clock reads $t^* = 0$, he would observe the $x = 800$ clock to be on the left side of him. His conclusion then is that the unstarred clocks are spaced less than the 800 feet he claims his own clocks are spaced.

Tachyons

Question 13-5:

> Use the snapshot diagram to demonstrate some causality difficulties that would arise if signals or objects could travel faster than light through a vacuum.

Recently a considerable amount of literature has appeared on the possible existence of objects (tachyons) traveling faster than the speed of light. (See Newton's article in *Science*, which contains a comprehensive bibliography of the earlier literature.[3]) Some controversy has arisen over what implications the existence of such objects would have on the validity of the special relativity theory, particularly with respect to questions of causality. These causality problems can be shown to occur whenever the speed of the object is greater than the speed of light through a vacuum. In our discussion, for simplicity we assume the speed of the tachyon to be infinite.

Refer to Figure 13-1. Suppose an observer in the unstarred frame at $x = 0$ has a cage containing a particle that becomes a tachyon when released from the cage. Assume he pushes a button at time $t = 0$, which initiates the release of the particle. The particle then travels toward the right with an infinite (or very nearly infinite) speed relative to the unstarred frame. It would reach $x = 800$ and $x = 1600$ at essentially $t = 0$, because a zero (or very nearly zero) time interval is required for the object to traverse the distance involved. Therefore, none of the clocks in either frame would change from their reading in Figure 13-1 during the time the tachyon was passing across the diagram.

If the starred observers recorded their observations on this moving object, they would get the following raw data for the sequence of events corresponding to the tachyon's location: it passes $x^* = 0$ at $t^* = 0$; it passes $x^* = 800$ at $t^* = -16/25$ μs; it passes $x^* = 1600$ feet at $t^* = -32/25$ μs; etc. How would the observers in the starred frame interpret these data? They would claim that the object was at $x^* = 1600$ feet before it was at $x^* = 800$ feet, and was at $x^* = 0$ later yet. Their conclusion would have to be that the object was traveling to the *left* at a speed of $800/(16/25) = 1250$ ft/μs.

Cause–Effect Difficulty

From the unstarred view the cause of the events was the pushing of the button that released the object. But from the starred view, the object was at $x^* = 1600$ or $x^* = 800$ *before* the button was pushed at $t^* = 0$. Thus the pushing of the button could not be interpreted, at least in the usual sense, as the "cause" of the release. The existence of such a tachyon would seem to raise embarrassing difficulties with regard to notions of cause and effect.

Before continuing our discussion of these cause–effect problems, let us compute the answer to the following question. According to the starred frame at time $t^* = -32/25\ \mu s$, where is the unstarred origin and what does the unstarred origin's clock read? In other words, given that the event has $x = 0$ and $t^* = -32/25\ \mu s$, what are the values of x^* and t? To answer this, we substitute $x = 0$, $t^* = -32/25$ into equation (13-1) and solve for t. This gives $t = -96/125\ \mu s$. Next we substitute $x = 0$, $t = -96/125$ into equation (13-2) and solve for x^*. This gives $x^* = 1024$ feet. So the event $x = 0$, $t = -96/125$ corresponds to $x^* = 1024$, $t^* = -32/25$.

Now let us determine the consequences of the result of this last calculation. According to the starred observers, at time $t^* = -32/25\ \mu s$, the cage at $x = 0$ is located at $x^* =$ 1024 feet and the clock on the cage reads $t = -96/125$. The tachyon wasn't released until the clock on the cage read $t = 0$, so the particle is in the cage. But at this same time $t^* = -32/25$ the starred observers know the object is outside the cage at $x^* = 1600$ feet and moving toward the left. Therefore, from the viewpoint of the starred frame, at the instant $t^* = -32/25\ \mu s$, the object would be inside the cage (at $x^* = 1024$) at the same time it was *outside* the cage (at $x^* = 1600$) traveling toward the cage. In fact, at any time before $t^* = 0$, there would be two objects, one inside and one outside the cage. At $t^* = 0$ the two objects apparently would "annihilate" each other and neither would exist after $t^* = 0$.

If the unstarred frame is allowed to have objects traveling with infinite speed relative to it, then the principle of relativity requires us to allow the starred frame also to have similar objects that travel with infinite speed relative to the starred frame. Suppose the $x = 0$ observer has a button which, when pushed, releases an object traveling to the right with an infinite speed relative to the unstarred frame.

Also suppose the $x^* = 1600$ observer has a button which, when pushed, releases an object traveling to the left with an infinite speed relative to the starred frame. The $x^* = 1600$ observer (in the starred frame) is given instructions to push his button *only* if and when he observes a faster-than-light particle passing by. The $x = 0$ observer is to push his button at $t = 0$ only if he has not observed a faster-than-light object passing by before $t = 0$. We can show that it is not possible for the $x = 0$ observer to follow his very simple instructions.

Causal Anomaly

Assume the $x = 0$ observer did not see any faster-than-light particles pass by his location before $t = 0$. Then, according to his instructions, he should push the button. If he pushes the button, then the object he releases will pass the $x^* = 1600$ observer at $t^* = -32/25$ μs. The $x^* = 1600$ observer, following his own instructions, should release his object at $t^* = -32/25$ μs. This means that the latter object would be at $x^* = 1024$ at $t^* = -32/25$ μs (since it is traveling to the left at infinite speed, according to the starred frame). But this latter event is the same as $x = 0$ at $t = -96/125$, according to our previous calculation. So the $x = 0$ observer should have seen this faster-than-light particle passing by before $t = 0$, and he should not have pushed the button at $t = 0$. We are led to conclude that there is no way the $x = 0$ observer can follow his instructions. This is referred to as a "causal anomaly," and it represents an insurmountable contradiction from the viewpoint of the $x = 0$ observer. A detailed discussion of the causal anomaly has been given by Rolnick.[4]

The controversy over whether or not tachyons might exist centers on whether the foregoing causal anomaly can be avoided by proposing some property for the tachyon that would not allow the preceding thought experiment to be performed, even in principle.

Some Speeds Greater Than c Are Possible

Question 13-6:

> Must we conclude from the foregoing discussion that speeds greater than c cannot occur or have no meaning?

The answer is definitely "*no*!" We can conclude only that signals cannot be sent faster than the speed of light c. It is possible to have patterns or disturbances traveling from place to place at a speed greater than c, but none of these can be used to send signals faster than c. As an example, suppose we place a collection of people side by side along the x-axis of the unstarred reference frame. We give them instructions to raise their right arms in unison at $t=0$. If they do this, we have manufactured an arm-raising "disturbance" that moves down the line of people at infinite speed, according to us; that is, the arm-raising occurs everywhere simultaneously. How would this appear to the starred observers? According to the discussion following Question 13-5, they would conclude that the arms farther right would raise first and later the arms farther left would begin to raise. They would observe the arm-raising disturbance to be moving toward the left at a speed of 1250 ft/μs, and therefore greater than c. However, neither reference frame could take advantage of this phenomenon to send signals faster than c. The cooperation of the arm-raising observers had to be prearranged, and no information by this method would be sent from one of the observers to another at a speed greater than c.

Electrons striking the face of an oscilloscope tube cause a flash of light to be emitted. If electrons strike different places sequentially, the light flashes come from different places at different times, and a pattern of light flashes moves across the screen. The speed of this pattern may very well be greater than c. But again, no observer at one spot on the screen could use this method to send a signal to another observer at a different spot on the screen.

Valid speeds greater than c can also arise in a different manner. If one object is traveling to the left at speed $(4/5)c$ with respect to us and another object is traveling to the right at a speed of $(4/5)c$ with respect to us, then their relative speed is $(8/5)c$, *according to us.* This merely means that the rate at which their separation is increasing (or decreasing if they are approaching each other) is $(8/5)c$. No object or signal is traveling faster than c. If we ask an observer traveling along with one of the objects to determine the relative velocity from his viewpoint, then he will give an answer of less than c.

Suggested practice problems to ponder in Appendix A: P-20; P-21.

References

1. Terrell, J., *Physical Review*, vol. 116, 1959, p. 1041.
2. Weisskopf, V.F., *Physics Today*, vol. 13, 1960, p. 24.
3. Newton, R.G., *Science*, vol. 167, 1970, p. 1569.
4. Rolnick, W.B., *Physical Review*, vol. 183, 1969, p. 1105.

CHAPTER 14

Miscellanea Involving Time Dilation

Some of the particles formed in nuclear reactions are unstable and have very short half-lives. For example, one such particle, the mu meson or muon, has a half-life of 2.2 μs, from the viewpoint of a reference frame with respect to which the muons are at rest. This means that if we start out with a large sample of muons at rest, then half of them should survive the first 2.2 μs without undergoing decay into an electron, a neutrino, and an antineutrino. However, often the muons are traveling at a high speed when they are formed.

Muons as Clocks

Question 14-1:

A sample of muons is traveling at a speed of $0.994c$ with respect to us. What should be the half-life of the moving muons from our viewpoint?

With respect to the rest frame of the muons, the half-life would be 2.2 μs. Therefore, half the muons would undergo decay while 2.2 μs ticked off on the muon rest frame's clocks. But, according to us, the muon rest frame's clocks are running slow by the time dilation factor. So the time we say really elapsed is

$$\frac{2.2}{\sqrt{1 - V^2/c^2}} = \frac{2.2}{\sqrt{1-(0.994)^2}}$$

A bit of arithmetic gives $\sqrt{1-(0.994)^2}$ equal to about 1/9. So the half-life, according to us, is $2.2/(1/9) = 2.2 \times 9 = 19.8$ μs.

The result computed in the foregoing has been experimentally verified.[1]

Mention was made in Chapter 8 of controversy surrounding the twin paradox. Some people were unable to accept the conclusion that the clock making the round trip should register less time than the clock left behind.

Experiment with Accelerated Clocks

Question 14-2:

> Suppose a beam of muons is traveling in a circle at a speed of $0.9965c$. If the time dilation factor does apply to these muon "clocks" even though they are accelerated (centripetal acceleration), then what value should we get for their half-life?

As in the preceding question, the half-life according to us should be $2.2/\sqrt{1 - V^2/c^2}$, and $\sqrt{1-(0.9965)^2} = 1/12$. So the half-life would be $2.2 \times 12 = 26$ μs.

Again, an experiment precisely of the type mentioned in Question 14-2 has been performed that confirms the correctness of the above result. See *Nature* for a report of the experiment by Farley, Bailey, and Picasso.[2]; also a very brief discussion of the experiment appeared in *Physics Today*.[3] In this experiment the muon clocks were making round trips and did "run slower" than did the muon clocks at rest. Thus there has been quantitative experimental confirmation for the correctness of the analysis carried out in Chapter 8 concerning the round trip to a star.

Question 14-3:

> An observer is looking through a telescope at a vibrating object moving with respect to the observer. What is the relation between the frequency of vibration as seen through the telescope and the frequency of vibration as measured by an observer at rest with respect to the vibrating object?

Suppose a clock is moving either away from or toward an observer watching the clock through a telescope. Let Δt be the true time interval from the viewpoint of the observer, and let Δt^* be the apparent time interval registered by the moving clock as seen through the telescope during the time interval Δt. According to equation (8-2) in Chapter 8, the formula relating Δt and Δt^* is

$$\Delta t = \frac{\Delta t^* (1 - V/c)}{\sqrt{1 - V^2/c^2}} \qquad (14\text{-}1)$$

where V (the speed of the clock) is chosen to be positive if the clock is approaching the observer with the telescope and negative if the clock is receding from the observer.

Suppose some object moving with the clock is seen to vibration n times during the time interval. The frequency (vibrations per second) of the vibrating object would be $n/\Delta t^*$ according to an observer in the rest frame of the vibrating object. So $n/\Delta t^* = f_0$, where f_0 is the rest frequency of the vibrating source (frequency as measured by an observer at rest with respect to the source). However, the observer looking through the telescope at the moving object would also see n vibrations, but they occur over a time interval Δt. The effective observed frequency is $f = n/\Delta t$ as determined by the observer as the telescope. We can use (14-1) to obtain the formula connecting f and f_0.

$$f = \frac{n}{\Delta t} = \frac{n}{\dfrac{\Delta t^*(1 - V/c)}{\sqrt{1 - V^2/c^2}}} = \left(\frac{n}{\Delta t^*}\right) \frac{\sqrt{1 - V^2/c^2}}{(1 - V/c)}$$

$$= f_0 \frac{\sqrt{1 - V^2/c^2}}{(1 - V/c)}$$

or

$$f = f_0 \frac{\sqrt{1 - V^2/c^2}}{(1 - V/c)} \tag{14-2}$$

An alternative form of (14-2) can be obtained by substituting $(1 + V/c)(1 - V/c)$ for $(1 - V^2/c^2)$. *The result* is

$$f = f_0 \frac{\sqrt{1 + V/c}}{\sqrt{1 - V/c}} \tag{14-3}$$

Remember, V is positive if the object is approaching the observer, and negative if it is receding from the observer.

Relativistic Doppler Effect

The frequency relation expressed in (14-2) or (14-3) is known as the equation for the relativistic Doppler effect. It is an important formula for interpreting some astronomical observations. The frequencies measured for the light emitted by hydrogen atoms on the most distant galaxies are found to be much less than the corresponding frequencies of the light emitted by hydrogen atoms at rest on the earth. The explanation proposed for this phenomenon is that the distant galaxies are receding from us. If the shift in one of the frequencies is known from measurements, equation (14-3) can be used to compute the speed V of the galaxy. Then this same V inserted in (14-3) should give the correct result for the shift in the other frequencies emitted by hydrogen. In fact, the same V should account for the shift in the frequencies emitted by *all* other atoms in the galaxy. The Doppler effect explanation is adequate to account for most of the known experimental data on the observed shifts in frequencies of the light from the galaxies, and is now accepted as the correct explanation. Very recently some observations on other astronomical objects have raised questions about whether other mechanisms might be responsible for the frequency shifts in some situations, and the debate continues.

Expanding Universe

We are part of what is apparently an "expanding universe," with the galaxies receding from each other. The galaxies at a distance of 8 billion light-years away from us are receding from us at about half the speed of light.

> *Suggested practice problem to ponder in Appendix A: P-8.*

References

1. Frisch and Smith, "Measurement of the Relativistic Time-Dilation Using Mu-Mesons," *American Journal of Physics*, vol. 31, May 1963, p. 342.
2. Farley, Bailey, and Picasso, *Nature*, vol. 217, 1968, p. 17.
3. *Physics Today*, vol. 25, Jan. 1972, p. 11.

CHAPTER 15

Where To from Here?

We have completed the modest mission of this book – that is, to discuss in depth the novel space–time concepts arising out of Einstein's principle of relativity. The disagreement on simultaneity, time dilation, the Lorentz contraction, and the relativistic addition law of velocities constitute a major upheaval in the old ideas.

But Einstein did not stop here with his theory. He insisted that *all* of physics should be compatible with these new space–time properties and the principle of relativity. The laws of physics must take exactly the same form in *every* inertial frame. If the contrary were true, then the different form of the laws in different inertial reference frames could be used as the basis for setting up a criterion for giving preference to one inertial frame over another. Einstein put all the known physics laws to a stringent test. He assumed, tentatively, that each one of the laws was true in *one* inertial reference frame. Then, using the Lorentz transformation equations (instead of the old Galilean transformation equations), he determined what form the law took from the viewpoint of a different reference frame. If the law in the new frame didn't look exactly the same as the law did in the original frame, he concluded that the law was wrong. And then he tried to discover a new law to replace the old one, usually by finding the simplest modification of the old law that would satisfy his test.

For example, one can show that if Newton's second law of motion ($F = ma$) is assumed to be true in an unstarred frame, the same form of the law ($F^* = m^*a^*$) automatically would be true in a starred reference frame *if the old Galilean transformation equations were valid.* However, if the Lorentz transformation rules are applied to $F = ma$, they do *not* yield $F^* = m^*a^*$. Therefore, Einstein concluded that Newton's second law was not correct and he succeeded in finding a correct revised form. Similarly, conservation of energy didn't satisfy his test unless mass was added to the list of forms of energy, with the amount of energy stored in every mass object equal to the mass times c^2; that is, $E = mc^2$. On this basis he also predicted that mass should be convertible into other forms of energy and that the mass of an object should increase as it travels faster. All his predictions were verified later.

The laws of electricity and magnetism as expressed by Maxwell in the mid-1800's turned out to be consistent with Einstein's new ideas, so Einstein didn't have to modify them.

Einstein ran into great difficulty in finding a law of gravitation that was in harmony with his principle of relativity and his new space–time concepts. Newton's law of gravitation had worked very nicely for explaining and predicting astronomical observations and was consistent with all other known gravitational phenomena. Nevertheless, it was not compatible with the principle of relativity and the Lorentz transformation. So it *must* be wrong, according to Einstein. Einstein searched for a minor modification of Newton's law of gravitation that would pass his test. None worked. One reason for the difficulty was that the spirit of the law was out of harmony with relativity theory. Newton's gravitational law is an example of an action-at-a-distance force law; that is, one mass object exerts a gravitational force *instantaneously* on another mass object somewhere else. The force varies inversely as the square of the distance between the masses. If the masses are separated by distance d, as shown in Figure 15-1, then m_1 will exert a gravitational force F on m_2, as indicated. Suppose I wiggled m_1 back and forth a small amount. Then d would change values, and immediately the magnitude of the force on m_2 would vary accordingly. If someone near m_2 had a sensitive device for detecting the force on m_2, I could signal him instantaneously by wiggling m_1. But such instantaneous signaling gives rise to causality problems, as discussed in Chapter 13. Consequently Einstein had to look for a completely new approach to a gravitational theory to replace Newton's theory.

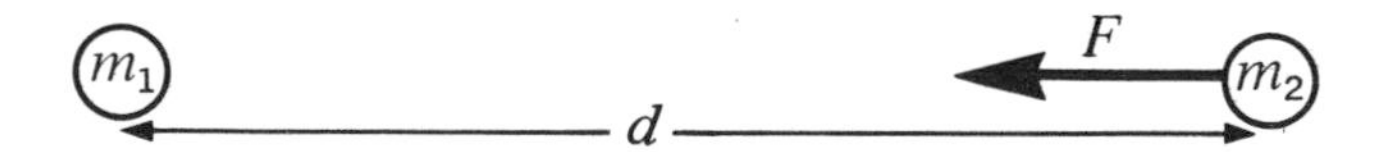

FIGURE 15-1:
Gravitational Force Between Two Masses, According to Newton

It took Einstein 10 years to put together a satisfactory gravitational theory, and to do this he had to bring in more new ideas on space and time in addition to those we have already discussed. His overall relativity theory, which includes his theory of gravitation, is called the "general theory of relativity." The theory of relativity without the extra features required to account for gravitation is known as the "special" or "restricted" theory of relativity.

It is hoped that the insight and understanding of the space–time concepts discussed in the earlier chapters of this book will be of benefit if the reader should decide to delve more deeply into the literature concerning the developments mentioned in this chapter. Any further discussion of these ideas here would be outside the intended scope of this book.

In Chapter 1 it was pointed out that quantum theory, the other half of what is known as modern physics, evolved during the early part of the 20th century, as did relativity theory. A major portion of quantum theory does not satisfy Einstein's test for a satisfactory theory, and physicists have not yet succeeded in making quantum theory and the restricted theory of relativity mutually compatible. And even less progress has been made in harmonizing quantum theory with the general theory of relativity. One of the most puzzling, frustrating, and exciting fundamental problems facing physicists at the present time is that of unifying these theories into a single, self-consistent, overall theory. Many brilliant men, including Einstein, have tried to accomplish this goal, but so far all have failed. The one who eventually succeeds will win fame equal to that of Galileo, Newton, and Einstein himself.

APPENDIX **A**

Practice Problems

(For answers see Appendix B)
Note: $c = 1000$ ft/μs

P-1. Suppose, as in Chapter 2, that the train is traveling to the right with respect to the tracks. But now assume that when A on the tracks is opposite the back of the train, the train clock there reads zero. And when B on the tracks is opposite the front of the train, the train clock there reads zero. (This means that these two events are simultaneous according to the *train* observers.) Suppose the track clock at A also reads zero when the back of the train is opposite it.

(a) According to the track observers, which event (A or B) was earlier?

(b) Draw a diagram showing the situation at zero time according to the train observers. (Show clocks on train and at A and B on tracks.)

(c) Draw a diagram showing the situation when the back of the train is opposite A, from the viewpoint of the track observers. (Show clocks on train and at A and B on tracks.)

(d) Draw a diagram showing the situation when the front of the train is opposite B, from the viewpoint of the track observers.

(e) According to the train observers, the distance between A and B on the tracks is exactly the same as the length of the train. [This should be apparent from your diagram for part (b), if you drew the diagram correctly.] According to the track observers, how does the length of the train compare with the distance between A and B on the tracks? [Hint: Refer to your diagrams in (c) and (d).]

P-2. A train goes by at a speed of $(3/5)c$ toward the right relative to the tracks. (See diagram.) When end 1 is opposite point A on the tracks and end 2 is opposite point B on the tracks, lightning strikes the two ends *simultaneously from the train observers' viewpoint.* According to the train observers, the length of the train is 60 meters.

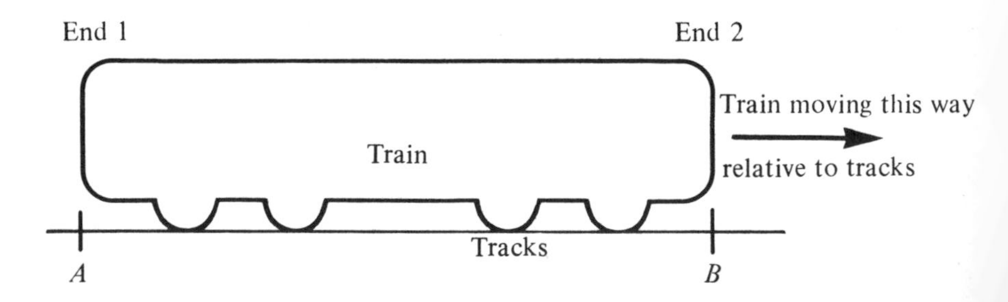

(a) Which flash occurred first from the track observers' viewpoint?

(b) How long is the moving train from the track observers' viewpoint?
(c) How far apart are points A and B on the tracks, from the track observers' viewpoint?
(d) How far apart are points A and B on the tracks, from the train observers' viewpoint?
(e) According to the train observers, how much time will tick off on one of the track clocks during the time 20 seconds elapse on the train clocks?
(f) A window on the side of the train has an area of 10 square feet according to the train observers. What is the area of the moving window from the viewpoint of the track observers?
(g) A bullet is shot toward the front of the train. It travels at a speed of $(1/3)c$ relative to the train. What speed is the bullet traveling relative to the tracks, from the track observers' viewpoint?
(h) A neutrino is traveling toward the front of the train at a speed of c (i.e., at the speed of light) according to the track observers. What is the speed of this neutrino from the viewpoint of the train?
(i) Suppose the train now comes to rest on the tracks. How long do the track observers say the train is?

P-3. A rocket goes in a straight line at a speed of $(4/5)c$ past me. My clock and the rocket's clock both read zero as they pass each other. (I am in an inertial reference frame.) When my clock reads 1 second, I send a light signal toward the rocket.

(a) What did the clock on the rocket read at the instant I sent the beam, according to me and my observers?

(b) What did the clock on the rocket read at the instant I sent the beam, according to the rocket observers?

(c) What will the clock on the rocket read when the light beam catches up to it?

(d) Now suppose the rocket keeps its speed of $(4/5)c$ but turns around and comes back to me. When it gets back, the rocket clock reads 100 seconds. What does my clock read when it gets back?

P-4. Consider three different inertial reference frames A, B, and C. According to reference frame B, reference frame A is traveling to the left with a speed of v meters/second, while reference frame C is traveling to the right at a speed of $2v$ meters/second. (v is a positive number.) The x-axes of all the frames are parallel to the left–right direction.

(a) According to A, is the speed of C relative to A greater than, less than, or equal to $3v$?

(b) According to B, is the speed of C greater than, less than, or equal to $3v$?

(c) A meter stick is at rest in B and is oriented parallel to the x-axis. Suppose all three reference frames measure the length of this meter stick while it is at rest in B.

(1) Which of the three frames gets the largest value for the length of the meter stick?

(2) Which of the three frames gets the smallest value for the length of the meter stick?

P-5. Reference frame O^* is moving to the right with a speed of $(4/5)c$ with respect to reference frame O. A right triangle made out of pieces of wood is at rest in O^*, and according to O^* the triangle has the dimensions shown in the diagram:

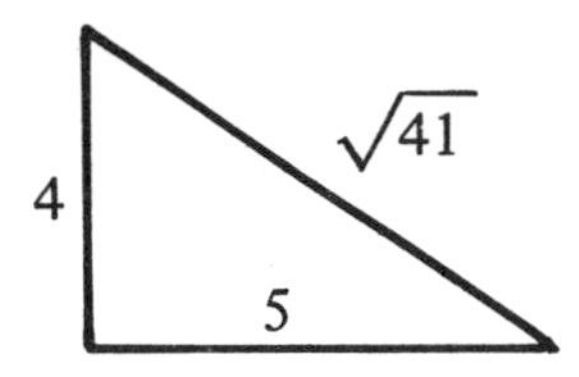

Draw the dimensions of this (moving) triangle from the viewpoint of reference frame O.

P-6. A starred clock goes by me at a constant velocity of $(4/5)c$. My clock and the starred clock both read zero as they pass each other. When my clock reads 1 μs, I suddenly start moving so that I am *gaining* on the starred clock at a relative speed of $(4/5)c$. [Thus the relative speed between my clock and the starred clock is $(4/5)c$ throughout the process.]

(a) What will my clock read when I catch up with the starred clock?

(b) What will the starred clock read when I catch up with it?

(c) In the foregoing process, what is the maximum separation distance between the two clocks, according to the starred clock's view?

(d) Describe the foregoing process in detail from the viewpoint of an observer on the starred clock.

P-7. A man is sitting in a train and he is pounding on a table with his fists. According to him, a time of 3 seconds elapses between successive pounds of his fist on the table. The train is traveling at a speed of

$(4/5)c$ relative to me. How long a time interval do I say elapses between successive pounds of the man's fist?

P-8. I am in an inertial reference frame. John is in a rocket that is traveling at a speed of $(3/5)c$ directly toward me, according to me. John is bouncing a ball up and down at the rate of 25 bounces per minute, according to John.

(a) What is the true bouncing rate of John's ball, according to me?

(b) Suppose I am watching through a telescope and see John bouncing the ball. What is the *apparent* bouncing rate of the ball as I watch it through the telescope?

(c) Now suppose John's rocket is going directly away from me at a speed of $(3/5)c$ according to me, and John is still bouncing the ball.

(1) What is the true bouncing rate of John's ball, according to me?

(2) What is the apparent bouncing rate of the ball as I watch it through a telescope?

P-9. George's reference frame is moving at a speed of $(3/5)c$ relative to me in the positive x direction. Two explosions on the x-axis were simultaneous and were 1000 feet apart, according to me.

(a) According to George, how far apart were the two explosions?

(b) According to George, how long a time elapsed between the explosions?

P-10. A train is traveling at a speed of $(2/3)c$ toward the right relative to me. A rocket is traveling at a speed of $(5/6)c$ toward the right relative to me.

(a) How fast is the rocket traveling relative to the train, according to the train observers?
(b) What is the relative speed of the rocket and the train, according to me?

P-11. Two identical clocks A and B were initially at rest side by side. Suddenly, when each read $t = 0$, *one* of the clocks started moving at a speed of $(4/5)c$ relative to the other in a circle. (The other clock did not change its original state of motion.) When the one clock completed the circle and got back to the other, it was found that clock A read 10 μs and clock B read 6 μs.
(a) Which clock went around the circle?
(b) How big was the circumference of the circle?

P-12. A father is 30 years old and his son is 10 years old. Suppose the father goes on a rocket trip at a constant speed relative to his son. How fast must the rocket go in order for both to be 60 years old when the father returns?

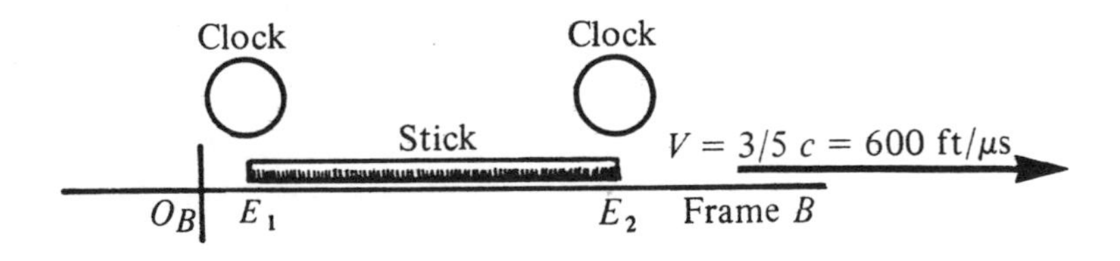

P-13. Reference frame B is moving to the right at a speed of $(3/5)c$ with respect to reference frame A. A stick with a rest length of 2000 feet is at rest in frame B. (See diagram.) A clock is sitting on each end of the stick (at E_1 and E_2), and the clocks are synchronized according to frame B.

(a) When the clock on end E_1 reads $t_B = 0$, a light beam starts out toward the right from E_1.

(1) What will be the reading on the clock on end E_2 when the light beam gets to it?

(2) According to frame A, how long does it take for the light beam to go between the two clocks at ends E_1 and E_2?

(3) According to frame A, how much time ticks off on each of the clocks on E_1 and E_2 during the time the light beam is going between the two clocks?

(4) According to frame B, how much time ticks off on each of frame A's clocks during the time the light beam is going between the two clocks on ends E_1 and E_2?

(b) According to frame A, what does the clock sitting on end E_2 read at the instant the clock on end E_1 reads $t_B = 0$?

(c) The stick is handed over to reference frame A, and it eventually comes to rest in frame A. How long is the stick now, according to frame A?

P-14. Two particles, A and B, are sitting at rest, with particle B a distance of 300 feet to the right of particle A. Suddenly particle B starts traveling toward the right at a constant speed of $(4/5)c$, and 1 μs later particle A also suddenly starts traveling toward the right at a constant speed of $(4/5)c$. At the end of 10 μs, both particles suddenly came back to rest simultaneously (according to us). How far apart are the two particles at the end of this process, according to us?

P-15. John's inertial frame is moving to the right at a speed of $(4/5)c$ with respect to George's inertial frame, according to George.

(a) An object is moving to the left at a speed of $(1/5)c$, according to John. At what fraction of the speed of light is the object traveling, according to George?

(b) Point B is 200 feet to the right of point A, according to John. An explosion occurs at each of these two points, and they are simultaneous according to John.

(1) What was the distance between the explosions, according to George?

(2) Which explosion occurred earliest, according to George?

(3) What was the magnitude of the time interval between the explosions, according to George?

(c) John is holding an object in his hands. George determines the shape of this (moving) object and finds it to be a sphere with a 12-foot diameter (while it is moving). Draw a diagram showing the shape and dimensions of this same object according to John.

(d) A mirror is placed at rest at a distance of 500 feet to the right of the origin of John's inertial frame, according to John. A beam of light leaves John's origin, strikes the mirror, and is reflected back to the origin.

(1) According to John, how long a time interval is required for this round trip for the light beam?

(2) According to George, how long a time interval is required for this entire trip?

(3) According to George, how long an interval elapsed between the time the beam left John's origin and when it struck the mirror?

(4) According to George, how long an interval elapsed between the time the beam struck the mirror and when it returned to John's origin?

P-16. Rocket A is traveling to the right at a speed of $(3/5)c$ relative to rocket B, according to rocket B. When the rockets pass each other, both of their clocks read zero. When B's clock reads 10 μs, he shoots a bullet toward rocket A. Rocket A measures the speed of the bullet to be $(3/5)c$ to the right relative to rocket A.

(a) What will rocket A's clock read when the bullet catches up to rocket A?

(b) According to rocket A, how far away was rocket B when he shot the bullet?

(c) According to rocket A, what did rocket A's clock read at the instant rocket B shot the bullet?

(d) According to rocket B, what did rocket A's clock read at the instant rocket B shot the bullet?

(e) According to rocket B, how far away was rocket A at the instant rocket B shot the bullet?

(f) How fast was the bullet traveling according to rocket B?

P-17. Reference frame F^* is moving to the right at a speed of $(3/5)c$ relative to reference frame F, according to frame F. Two clocks, A and B, are at rest in F^* and synchronized according to F^*. Clock B is 1000 feet to the right of clock A, according to F^*.

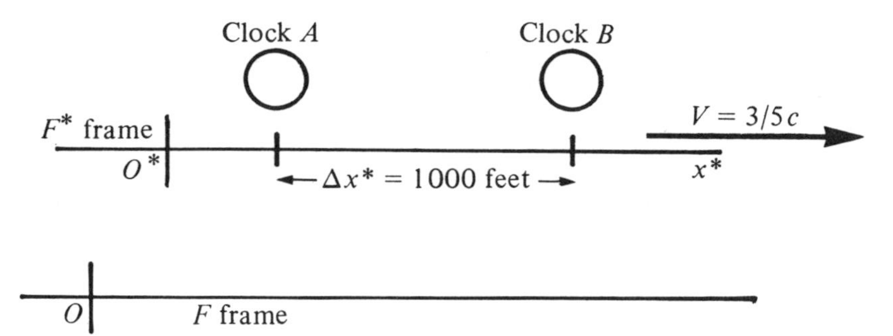

Diagram from F viewpoint

(a) How far apart are the two clocks, according to F?

(b) According to F, how much time elapses while clock A ticks off 10 μs?

(c) These two clocks are not synchronized, according to F.

 (1) At a given instant according to F, which clock reads the smaller value of time?

 (2) At a given instant according to F, what is the magnitude of the difference between the readings on clocks A and B?

(d) (1) When clock A read 1.0 μs a beam of light was sent toward clock B from A. What will clock B read when the beam reaches it?

 (2) According to F, how long an interval elapsed between the time the beam left clock A and when it reached clock B?

(e) A bullet was shot from clock B toward clock A. According to F^*, (5/4) μs elapsed while the bullet traveled between the two clocks.

 (1) How fast was the bullet traveling according to F^*?

 (2) How fast was the bullet traveling according to F?

(f) When clock A read 10 μs it suddenly came to rest in frame F. When clock B read 10 μs it also suddenly came to rest in frame F.

(1) Which clock stopped moving first, according to frame F?

(2) After this happened (both clocks now at rest in frame F), how far apart were the two clocks, according to frame F?

(3) After this happened how far apart were the two clocks, according to frame F^*?

P-18. Consider the same two reference frames, F and F^*, as shown in the preceding problem. Suppose George is at the origin of frame F^* and John is at the origin of frame F. Their clocks both read zero when they pass each other.

(a) According to George, how far away is John when John's clock reads 20 μs?

(b) According to John, how far away is George when John's clock reads 20 μs?

(c) According to George, what does George's clock read when John's clock reads 20 μs?

(d) According to John, what does George's clock read when John's clock reads 20 μs?

P-19. A flat, horizontal pan of dough speeds toward the right under a circular cookie cutter 10 cm in diameter at a speed of $(4/5)c$. A baker, holding the cutter perfectly horizontal, stamps the dough with lightning speed. (Assume he raises the cutter again so quickly that nothing gets squashed or stuck in the cutter.) The baker picks up one of these cookies and finds that it is not circular.

(a) What are the dimensions of the cookie (when it is at rest relative to the baker)?

(b) Justify your answer from the viewpoint of the frame of reference of the baker.

(c) Justify your answer from the viewpoint of the frame of reference of the dough.

P-20. Use the addition law of velocities to find v^* when v approaches infinity. Apply your result to confirm the value of v^* obtained in Question 13-5 in Chapter 13 for the case of a signal with v approaching infinity.

P-21. Consider a situation similar to that in Chapter 13. However, now assume the starred frame is moving to the right at $(3/5)c$ with respect to the unstarred frame. Suppose clocks are spaced at 600-foot intervals along the x-axes in both the starred and unstarred frames.

(a) Draw a diagram similar to Figure 13-1 showing the situation at the instant $t = 0$ from the unstarred view.

(b) Draw a diagram showing the situation at the instant $t = 1.0\ \mu s$ from the unstarred view.

APPENDIX B

Answers to Practice Problems

(For problems see Appendix A)

P-1. (a) The event at A is the earlier event.

(b)

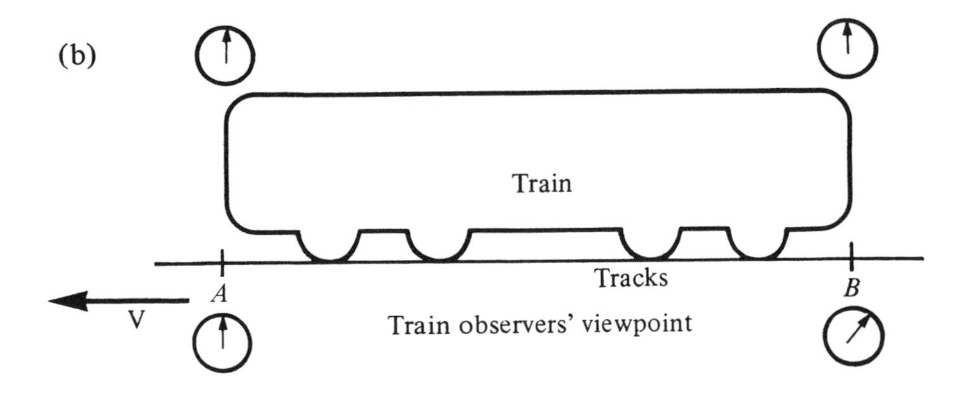

(c)

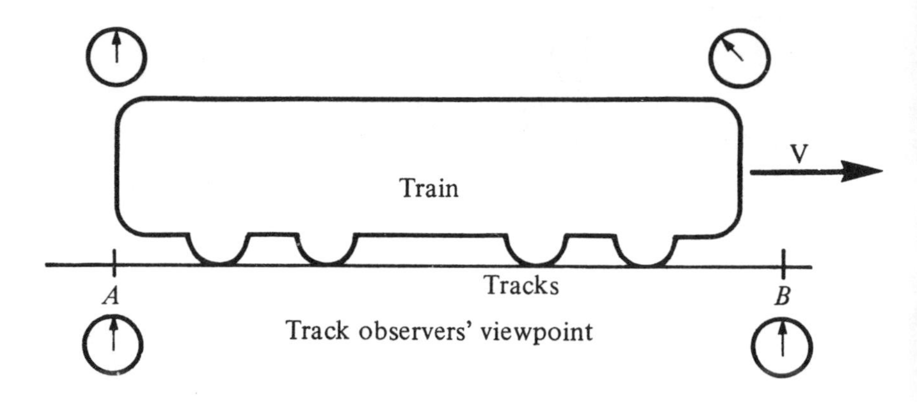

(d)

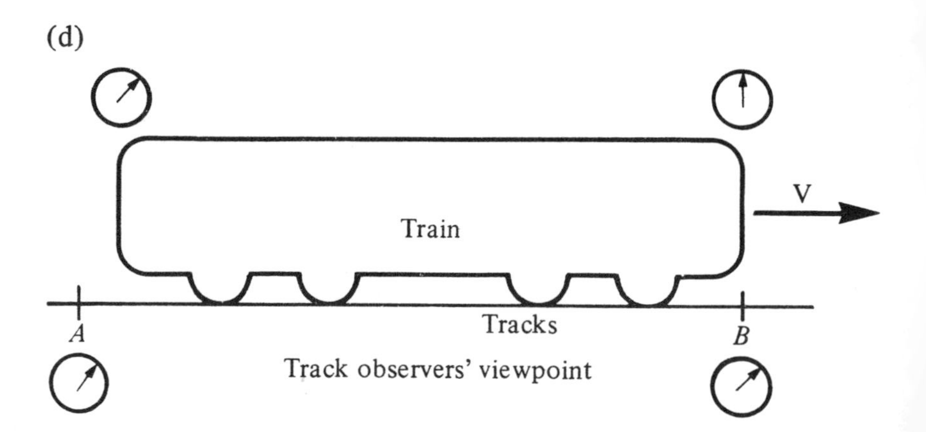

(e) From the diagram in (c) or (d) we see that the train is shorter than the distance between A and B, according to the track observers.

P-2. (a) A. (b) 48 meters. (c) 75 meters. (d) 60 meters. (e) 16 seconds. (f) 8 square feet. (g) $(7/9)c$. (h) c. (i) 60 meters.

P-3. (a) 3/5 second. (b) 5/3 second. (c) 3 seconds. (d) 167 seconds.

P-4. (a) Less than $3v$. (b) Equal to $3v$. (c) (1) B; (2) C.

P-5.

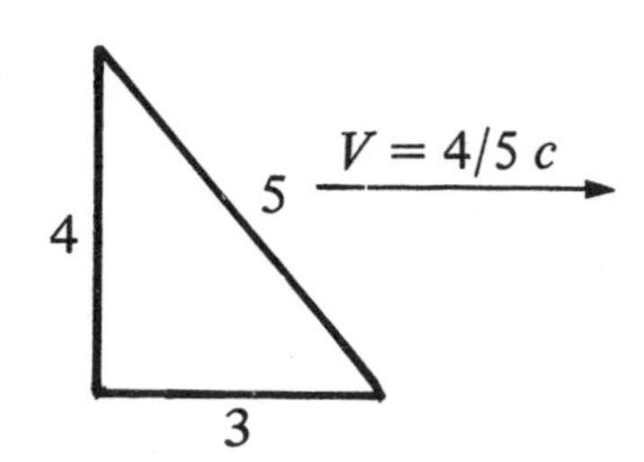

P-6. (a) 2 μs. (b) 3.33 μs. (c) 1333 feet. (d) The unstarred clock goes by at a speed of $(4/5)c$. It travels out for 5/3 μs and then turns around and comes back with the same speed of $(4/5)c$.

P-7. 5 seconds.

P-8. (a) 20 bounces per minute. (b) 50 bounces per minute. (c) 12.5 bounces per minute.

P-9. (a) 1250 feet. (b) 0.75 μs.

P-10. (a) $(3/8)c$. (b) $(1/6)c$.

P-11. (a) B. (b) 8000 feet.

P-12. $(4/5)c$.

P-13. (a) (1) 2 μs; (2) 4 μs; (3) 3.2 μs; (4) 1.6 μs. (b) -1.2 μs. (c) 2000 feet.

P-14. 1100 feet (no relativistic effect is involved in this problem).

P-15. (a) $(5/7)c$. (b) (1) 333.3 feet; (2) at point A; (3) (4/15) μs.

(c)

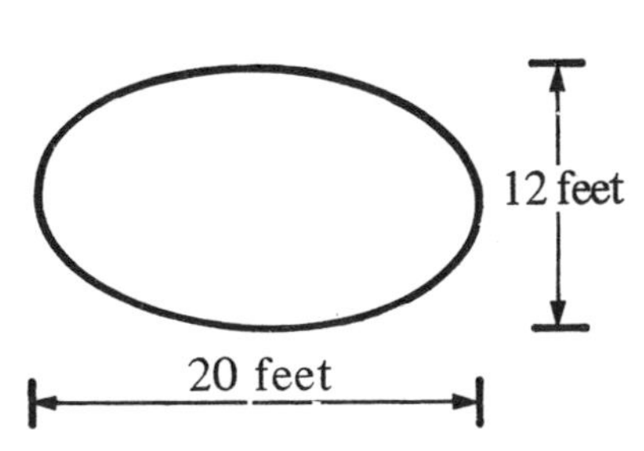

(d) (1) 1 μs; (2) 1.67 μs; (3) 1.5 μs; (4) 0.17 μs.

P-16. (a) 25 μs. (b) 7500 feet. (c) 12.5 μs. (d) 8 μs. (e) 6000 feet. (f) 882.35 ft/μs.

P-17. (a) 800 feet. (b) 12.5 μs. (c) (1) clock B; (2) 1.6 μs. (d) (1) 2 μs; (2) 2 μs. (e) (1) 800 ft/μs; (2) 384.6 ft/μs. (f) (1) A; (2) 1250 feet; (3) 1000 feet.

P-18. (a) 15,000 feet. (b) 12,000 feet. (c) 25 μs. (d) 16 μs.

P-19. (a)

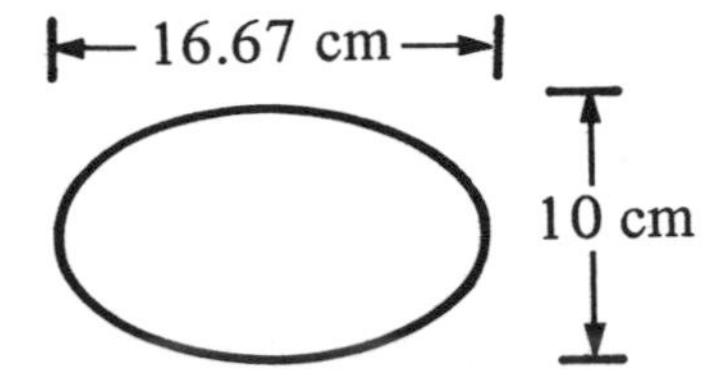

(b) The moving cookie was circular, but its rest shape would be as shown in (a) because of the Lorentz contraction.

(c) The cookie cutter, because it is moving, would have the shape in the diagram below. But the front of the cutter cut the dough at a different time than the back of the cutter, giving a final shape of the cookie as shown in (a) above.

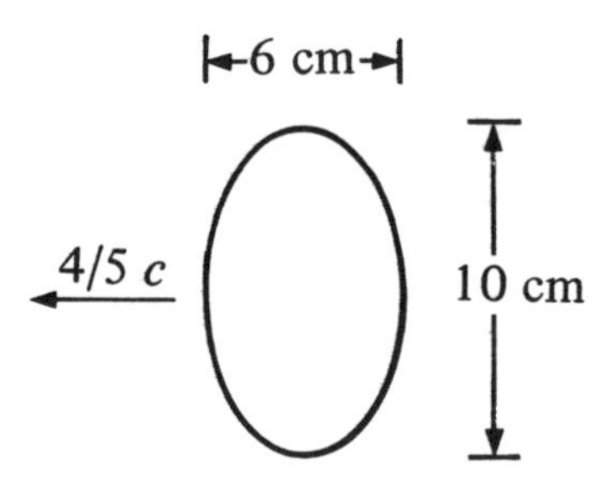

P-20. $v^* = \dfrac{v - V}{1 - \dfrac{vV}{c^2}}$

Now divide numerator and denominator by v. This gives

$$v^* = \frac{1 - V/v}{\dfrac{1}{v} - \dfrac{V}{c^2}}$$

As $v \to \infty$, then $V/v \to 0$ and $1/v \to 0$.
So for infinite v:

$$v^* = \frac{1}{-V/c^2} = -\frac{c^2}{V} = -\left(\frac{c}{V}\right)c$$

In Question 13-5, $V = (4/5)c$. So $v = -\frac{5}{4}c = -1250$ ft/μs. This agrees with the value obtained in Question 13-5.

P-21. *see pages 154 and 155.*

P-21.

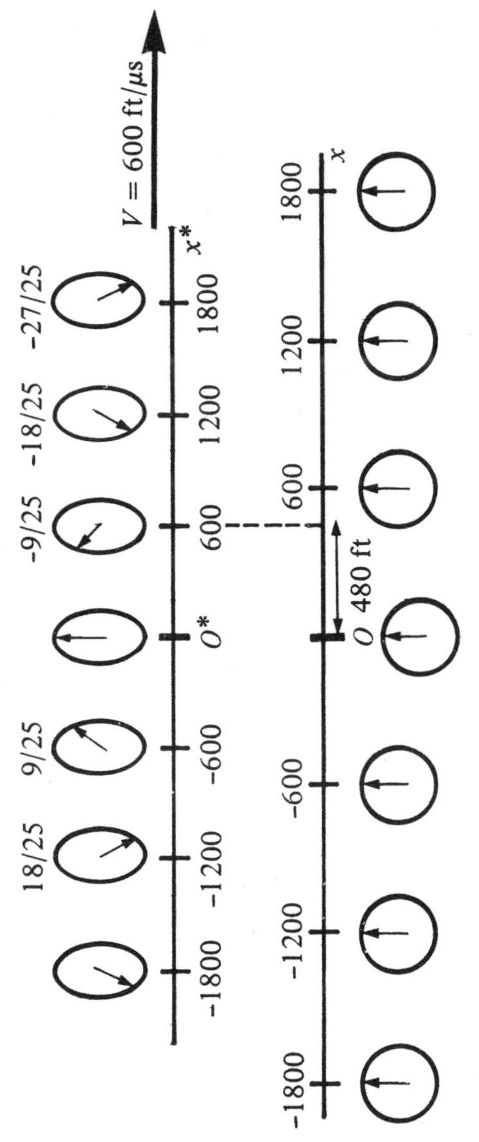

(a) Situation at $t = 0$ from the unstarred view.

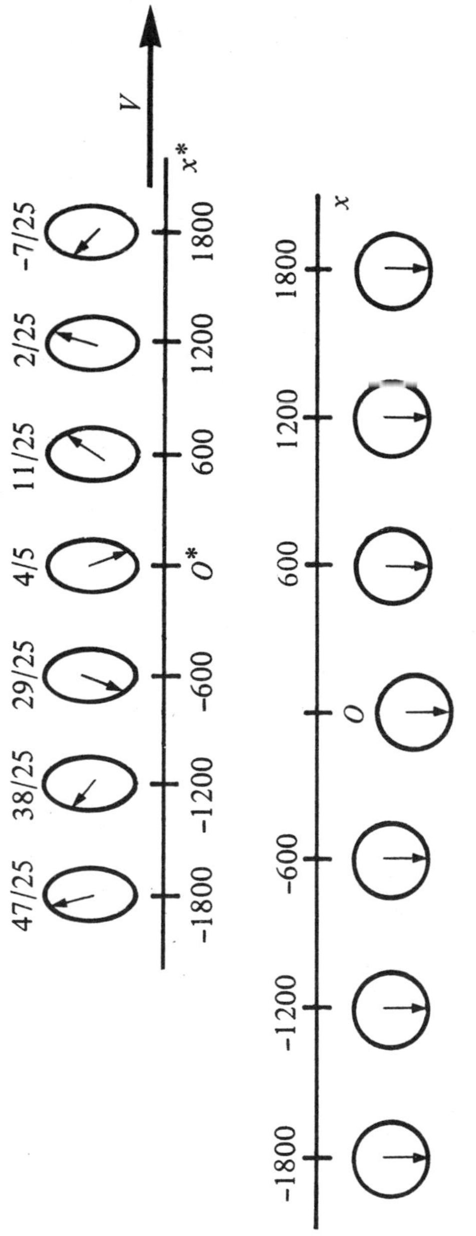

(b) Situation at $t = 1\ \mu s$ from the unstarred view.

Suggested Reading

The following is a partial list. It gives a representative sampling of elementary presentations of relativity theory.

Category A:
Those books which utilize even less mathematics than does this book.

Baker, A. *Modern Physics and Antiphysics.* Reading: Addison-Wesley, 1970. Chapters 4 through 9 give a very readable elementary account of the historical background of the concepts of relativity, along with a detailed discussion of simultaneity and the twin paradox. Other chapters present an excellent non-mathematical introduction to the concepts of quantum theory. I recommend it highly.

Einstein, A. and Infeld, L. *The Evolution of Physics.* New York: Simon and Schuster, 1951. This book concentrates on showing how the concepts of modern physics evolved to their current status. It is entirely descriptive and uses no mathematics. The book reflects the elegance and clarity of Einstein's mental processes.

Gamow, G. *Mr. Tompkins in Wonderland.* Cambridge: Cambridge University Press, 1940. This little book has become a classic because of its novel method of presenting ideas of modern physics to the layman. Mr. Tompkins attends three lectures on concepts of relativity and quantum theory. (These lectures are appendices of the book). After he hears the lectures, he has dreams in which he finds himself in strange universes. For example, one of his dreams finds him wandering about in a universe in which the speed of light is only about 10 miles per hour, so even bicyclists exhibit significant amounts of Lorentz contraction and time dilation.

Trefil, J.S. *Physics as a Liberal Art.* New York: Pergamon Press, 1978. This is a more recent text in the same spirit as the book by Einstein and Infeld described above. The theory of relativity is covered in Chapter IX. Also, see page 236 at the end of Chapter IX for some useful comments concerning additional suggested readings.

Category B:
Those books, or sections of books, which require about the same mathematics background as does this book.
Many recent general physics texts not using calculus would fall in this category.
Three representative examples are listed here.

Giancoli, D.C. *Physics.* Englewood Cliffs: Prentice-Hall, Inc., 1980. Concepts of special relativity are treated in Chapter 24, with a useful additional reading list at the end of the chapter.

Hewitt, P.G. *Conceptual Physics.* Boston: Little, Brown and Company, 1981. Now in its fourth edition, this is an unusually clear and readable general physics text. Chapter 8 is an excellent introduction to relativity and astrophysics.

Pasachoff, J.M. and Kutner, M.L. *Invitation to Physics.* New York: W.W. Norton and Company, 1981. Chapter 8 is on special relativity. This book is noted for its abundance of excellent line drawings, photographs, and illustrations.

Category C:
Those books, or sections of books, which utilize somewhat more advanced mathematics than does this book. Most calculus-based general physics and introductory modern physics texts would fall in this category.

French, A.P. *Special Relativity.* New York: W.W. Norton and Company, 1968.

Resnick, R. *Introduction to Special Relativity.* New York: John Wiley & Sons, Inc., 1968. Both of these are fairly comprehensive intermediate level texts, using some trigonometry and elementary calculus.

Gautreau, R. and Savin, W. *Schaum's Outline of Theory and Problems of Modern Physics.* New York: McGraw-Hill Book Company, 1978. The first nine chapters cover the special theory of relativity. It is most useful as a source of a multitude of solved problems.

Kaczer, C. *Introduction to the Special Theory of Relativity.* Englewood Cliffs: Prentice-Hall, Inc., 1967. Excellent comprehensive discussions of simultaneity, causality, doppler effect, and the twin paradox are presented, along with applications of relativity theory to dynamics and electrodynamics.

Kim, S.K. and Strait, E.N. *Modern Physics for Scientists and Engineers.* New York: MacMillan Publishing Company, 1978. The first three chapters are devoted to relativity theory. This is an examples of an excellent calculus-based modern physics text.

Shadowitz, A. *Special Relativity.* Philadelphia: W.B. Saunders Company, 1965. Although the last few chapters use some advanced mathematics, the earlier chapters are reasonably elementary. Chapter 2 discusses the uses of diagrams (Brehme, Loedel and Minkowski) to give schematic representations of relativistic effects.

Index